ESSENTIAL ORGANIC CHEMISTRY FOR BEGINNERS

Girija Shankar Singh | Ranjit Singh

INDIA • SINGAPORE • MALAYSIA

ISBN 979-8-89186-546-4

CONTENTS

CONTENTS IN CHAPTERS

PREFACE

Organic chemistry is considered a problematic subject by students across the globe. There are many reasons behind this. One among them is tough English language of the books written by native English speakers. Also, most international books include topics that are not required at lower levels. Even though there are many textbooks on basic organic chemistry we felt that there is need of a book on essential organic chemistry with the motto "make chemistry easy and simple". The present title is indeed essential organic chemistry that is required for beginners written in very simple language.

This book covers functional group chemistry of organic halogen compounds, alcohols, phenols, ethers, carbonyl compounds, and amines besides hydrocarbons in 12 chapters. The first two chapters before hydrocarbons are on introduction of organic compounds, structure and bonding, and acids and bases. Each chapter has 20-25 pages subdivided into structure, nomenclature, physical and chemical properties. The mechanism of reactions is described in a simplistic manner. At the end of each chapter, there is a summary of the chapter followed by exercises. There are some solved questions also within the chapters.

Like many other Indian books, it does not aim at a syllabus of any board or university, but it will suit to the students at high school, and pre-university students of India, and B. Sc. levels 1 and 2 students at universities across the world that have 4-years' university degree in chemistry.

This book is based on our lecture notes used for teaching introductory organic chemistry to beginners. The lecture notes, in turn, were prepared from some finest international books in the area that include i) Organic Chemistry – A short Course by Hart, Craine, and Hart, ii) Essential Organic Chemistry by Bruice, and iii) Organic Chemistry by McMurry. We are grateful to these renowned scientists and authors.

We are also grateful to late Prof. J. Ahmad, late Mr. I. Meulenberg, and Dr. G. Kumar, who coauthored with one of us (GSS) the book titled "General Chemistry". The chapter on chemistry of carbon compounds of this book forms the basis for chapters on alkanes, alkenes, and alkynes of the present book.

Since this is the first edition of the book, there may be errors here and there. We would be grateful to the authors if they bring in such errors or any suggestion for further improvement to our notice.

Girija Shankar Singh

Professor of Chemistry

University of Botswana, Gaborone, Botswana.

Ranjit Singh

Associate Professor of Chemistry

S. S. College, Sasaram (A constituent unit of V. K. S. University, Ara, Bihar, India).

1 CARBON COMPOUNDS: INTRODUCTION, CLASSIFICATION, STRUCTURAL FEATURES

1.1 What is Organic Chemistry?

The modern definition of organic chemistry states that it is *the branch of chemistry, which deals with the chemistry of carbon compounds*. Thus, all the organic compounds have carbon as a central atom. This definition has a broad scope. It accounts for both synthetic compounds and the compounds obtained from natural resources – plants, animals, marines, and microorganisms. However, carbon dioxide, carbon monoxide, carbonates and bicarbonates are excluded from the class of organic compounds because they greatly resemble inorganic compounds in properties.

1.2 A Brief Historical Development

The word organic is derived from the Greek word *organ* meaning *life*. It is interesting to know how a word that is signifying "life" relates to carbon-based materials. In ancient times, the curiosity of chemists about living beings led them to study the substances obtained from animals and plants. It soon became clear that compounds obtained from the living beings were different from those obtained from the non-living things such as minerals. Most compounds from animals and plants were made up of carbon, hydrogen, oxygen, nitrogen, and sometimes sulfur and phosphorus. Carbon was present in all the compounds. The chemists thus believed that these compounds are only formed inside the living beings and that is why the term "organic" was used to describe such compounds. This belief gave birth to the *vital force theory*. This theory postulated that living beings had some "vital force" that was responsible for the synthesis of carbon-based materials.

This belief was disproved in 1828 by a German chemist Friedrich Wohler, then only 28 years old, who prepared an organic compound urea, a constituent of urine and now well-known fertilizer, in the laboratory by heating ammonium cyanate, a compound occurring in minerals.

$$\underset{\text{ammonium cyanate}}{NH_4CNO} \xrightarrow{\text{heat}} \underset{\text{urea}}{H_2NCONH_2}$$

The synthesis of urea led to the development of a new era in the study of organic compounds. Chemists started investigating methods to synthesize other compounds isolated from plants and animals. Synthesis of compounds was necessary for the study of their properties because many times the amounts extracted from plants and animals were too small to study their properties. Very soon chemists were able to synthesize thousands of organic compounds. One might be led to think that the word "organic" is now a misnomer because its origin is not limited to living beings. This is, however, not true. The use of this word is justified in the sense that carbon compounds form the basis of life.

The major constituents of living beings - amino acids, carbohydrates, fat, nucleic acids, and hormones are organic compounds. A simple example of a carbohydrate, an amino acid and a fat is shown below (**Figure 1.1**). Almost all the reactions in plants and animals involve organic compounds.

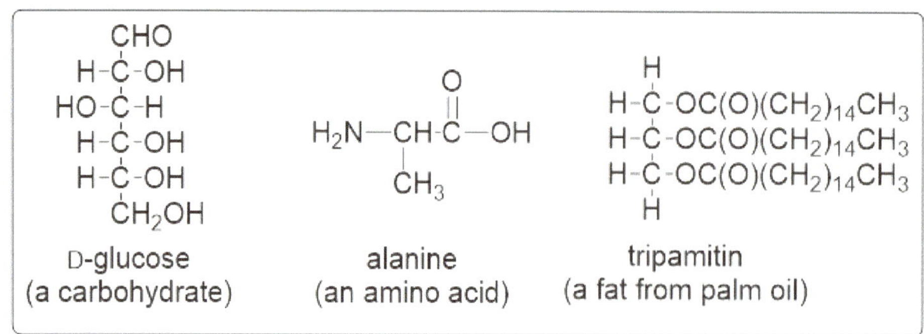

Fig. 1.1: *Example of a carbohydrate, an amino acid, and a fat.*

1.3 Organic Chemistry in Everyday Life

Organic chemistry is related to our day-to-day life more than any other branch of science. The paper of our books, the wood of furniture, our clothing, most of the medicines, gasoline, plastic and perfumes are all organic compounds. Some common examples of organic compounds (**Figure 1.2**) include acetic acid that is the principal constituent of vinegar, is used as preservative in pickles. You must be familiar with spirit that is used as antiseptic. Spirit is an alcohol, ethanol that is also used in preparation of many alcoholic beverages. Diethyl ether, commonly known as ether, besides being used as solvent in chemical laboratories, was in use as a general anesthetic compound. Nowadays its use is rather limited due to side-effects. Acetylsalicylic acid tablets are sold in pharmacies with the name aspirin as a painkiller and also as a blood thinner. A halogen containing alkene such as vinyl chloride is used for preparation of polyvinyl chloride (PVC) which is used as floor tiles or sheets. DDT (dichlorodiphenyltrichloroethane), a halogen derivative of hydrocarbon, is a well-known insecticide, while urea is a well-known fertilizer.

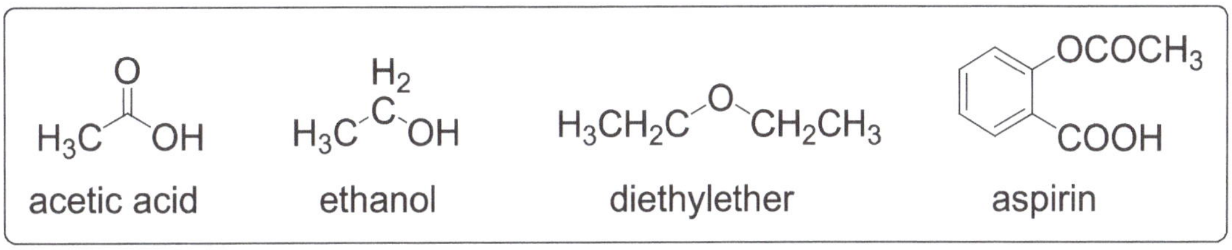

H_3C–C(=O)–OH — acetic acid
H_3C–CH_2–OH — ethanol
H_3CH_2C–O–CH_2CH_3 — diethylether
C_6H_4($OCOCH_3$)(COOH) — aspirin

Fig. 1.2: *Some common organic compounds.*

1.4 A Simple Classification of Organic Compounds

Millions of organic compounds are known to date. Their study would be extremely complex if they were not grouped together based on the types of atoms in them and their structural features. A simplistic approach is to put all the organic compounds into two groups, one with compounds containing only carbon and hydrogen atoms called hydrocarbons and the other comprising compounds containing other atoms as well besides carbon and hydrogen. Both groups can be further divided into several classes based on structural features such as the presence of carbon-carbon double or triple bond, presence of ring, and the presence of certain functional groups, which impart chemical properties to a compound. A brief classification of the main organic compounds is shown below (**Figure 1.3**). For details about important classes of compounds containing other atoms besides C and H, see functional groups.

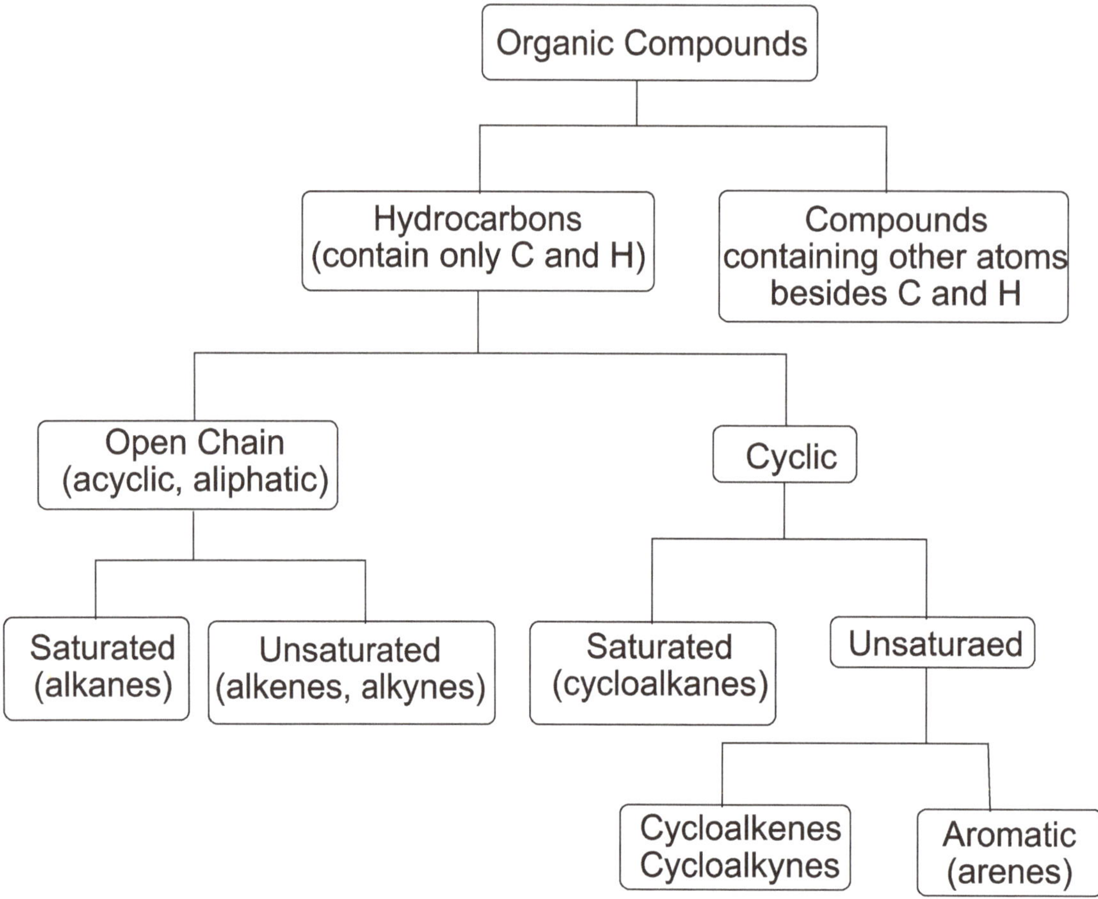

Fig. 1.3: *A simple classification of organic compounds*

1.5 Some Important Features of Organic Compounds

Besides having carbon in common, organic compounds have other characteristics, some of which are listed below.

a. Functional Groups

A **functional group** can be an atom or a group of atoms containing at least one **heteroatom** (atom other than carbon and hydrogen) present in the compound and is responsible for the properties of the compound. The common heteroatoms of the functional groups are oxygen, nitrogen, sulfur, and halogens. Some chemists treat double and triple bonds also as functional groups because their presence in the compound gives specific properties to the compounds. Compounds with the same functional group display similar chemical properties. Thus, the presence of functional groups forms the basis for classification of organic compounds containing atoms other than carbon and hydrogen atoms. A list of some important classes of organic compounds other than hydrocarbons, the functional group present in them, and simple examples are given in table below (**Table 1.1**). At this point of study, however, there is no need for you to memorize or get familiarized with all the functional groups. It is better for you to refer to Table 1 when you see organic products other than hydrocarbon during study on reactions of hydrocarbons.

Table 1.1: *Some Important Classes of Organic Compounds, Functional Groups, and Simple Examples.*

Functional groups	**Class**	**Examples**
O-H (hydroxyl)	alcohols, phenols	methyl alcohol or methanol: CH_3OH ethyl alcohol or ethanol: CH_3CH_2OH 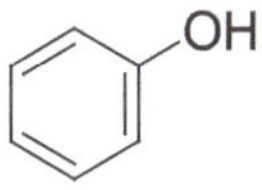 OH Phenol
-C-O-C-	ethers	dimethyl ether: CH_3-O-CH_3 ethyl methyl ether: CH_3-O-CH_2CH_3
R-CH=O R = alkyl, aryl.	carboxaldehydes or simply aldehydes	methanal or formaldehyde: H-CH=O ethanal or acetaldehyde: CH_3CH=O propanal: CH_3CH_2CH=O
R_2C=O R = alkyl, aryl	ketones	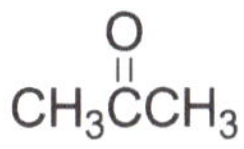 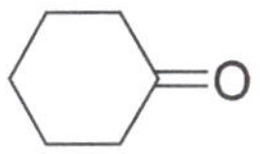 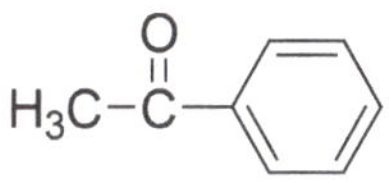 $CH_3\overset{O}{\overset{\|}{C}}CH_3$ (cyclohexane ring)=O $H_3C-\overset{O}{\overset{\|}{C}}-$(benzene ring) acetone cyclohexanone acetophenone

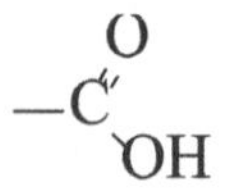 —C(=O)OH	carboxylic acids	methanoic acid or formic acid: HCOOH ethanoic acid or acetic acid: CH_3COOH COOH COOH Cyclohexane- benzoic acid carboxylic acid
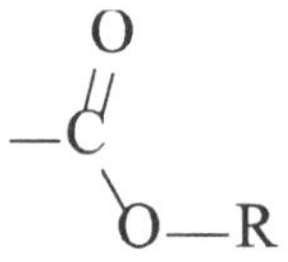 —C(=O)O—R	esters R = alkyl, aryl	$CH_3\overset{O}{\overset{\parallel}{C}}OCH_3$ $CH_3\overset{O}{\overset{\parallel}{C}}OCH_2CH_3$ $CH_3CH_2\overset{O}{\overset{\parallel}{C}}OCH_3$ methyl acetate ethyl acetate methyl propanoate
$-NH_2$	amines	methyl amine or methanamine: CH_3NH_2 ethyl amine or ethanamine: $CH_3CH_2NH_2$ isopropyl amine: $(CH_3)_2CHNH_2$
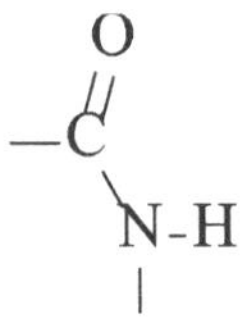 —C(=O)N-H	amides	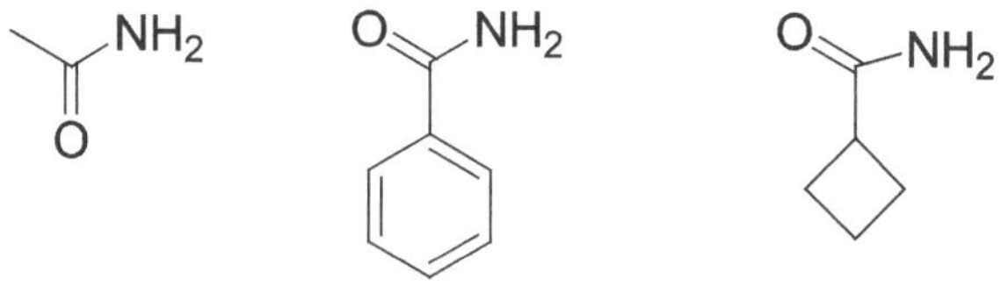 NH_2 O NH_2 O NH_2 O ethanamide benzamide cyclobutane-carboxamide

In the above list, esters and amides are treated as derivatives of carboxylic acids because they are prepared by the reactions of carboxylic acids with alcohols and amines, respectively. Some other important carboxylic acid derivatives are carboxylic acid halides, commonly called acyl halides, and carboxylic acid anhydrides.

During study on chemical properties of the compounds you will also encounter some other functional groups (**Figure 1.4**) such as nitro ($-NO_2$), nitrile or cyanide

(-CN), and sulfonic acid (-SO_3H) groups. Furthermore, like –OH and R-O-R, -SH (thiols) and R-S-R (thioethers or sulfides) are also common organic compounds. Organic compounds containing halogen atoms also constitute an important class of organic compounds and you will study their formation and reactions later in this book.

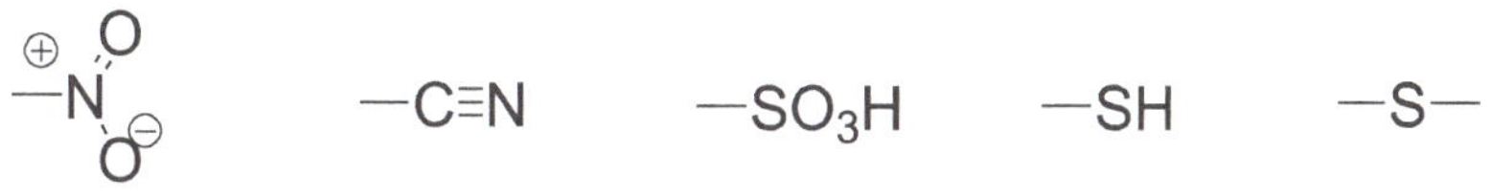

$-N^{\oplus}(=O)O^{\ominus}$	$-C\equiv N$	$-SO_3H$	$-SH$	$-S-$
nitro	nitrile	sulfonic acid	thiol	thioether

Fig. 1.4: *Some other functional groups.*

Problem: Identify the functional groups in the following molecule from left to right.

A. O N H OH O

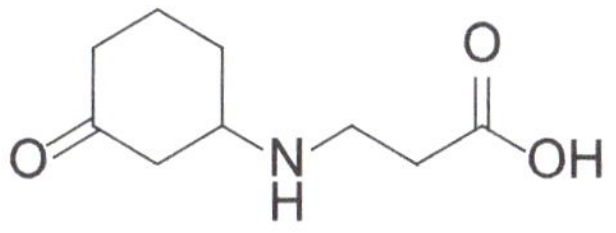

B. O O O NH$_2$ O

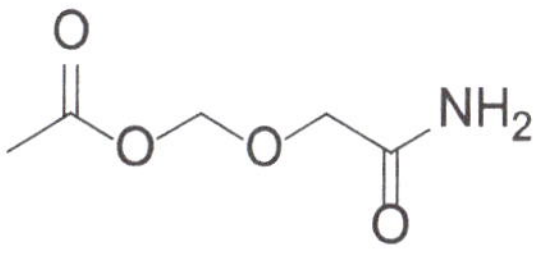

Solution: A. ketone, amine, carboxylic acid. B. carboxylic ester, ether, amide

b. Homologous series

Compounds of a series class, that show similar chemical properties, can be arranged in order of increasing molecular weight in a series. Each member of the series differs in molecular formula from the member above or below by CH_2. The series of compounds so arranged is called homologous series (see Table 2.1 for an example of homologous series of alkanes).

c. Isomerism

Isomerism is a very common property of organic compounds. There are many compounds known to have the same molecular formula but different structural formula meaning different connectivity of atoms in the formula or different orientation of atoms or groups in space (**Figure 1.5**). The compounds with identical structural formula show different properties due to different orientation of groups attached to certain carbon atoms in the space. The compounds with the same molecular formula but different structural formula are called *isomers*, and the property *isomerism*. Isomers with different connectivity of atoms in their molecules are called *structural isomers* or *constitutional isomers* and isomers with identical atom connectivity but different spatial arrangement of atoms or groups in their molecules are called *stereoisomers*. You will study in detail about different types of isomers in relevant chapters.

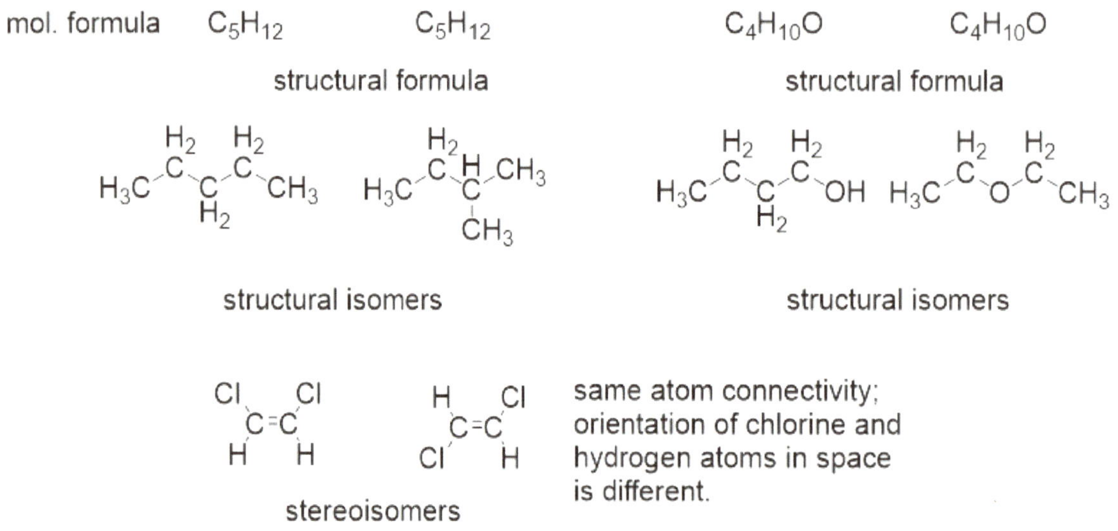

mol. formula C_5H_{12} C_5H_{12} $C_4H_{10}O$ $C_4H_{10}O$

structural formula

$H_3C-CH_2-CH_2-CH_2-CH_3$ $H_3C-CH_2-CH(CH_3)-CH_3$ $H_3C-CH_2-CH_2-CH_2-OH$ $H_3C-CH_2-O-CH_2-CH_3$

structural isomers

Cl(H)C=C(Cl)H   H(Cl)C=C(Cl)H

stereoisomers

same atom connectivity; orientation of chlorine and hydrogen atoms in space is different.

Fig. 1.5: *Some structural and stereoisomers.*

d. Catenation

Catenation is the ability of organic compounds to form long chains and large rings. You will see in studies ahead that carbon atom in organic compounds has remarkable ability to form long chains by covalent bonding among themselves and hence organic compounds can exist as long chains or large rings. There is almost no limit of chain length, and some compounds may contain over a hundred carbon atoms in a chain. The carbon chains might be straight (meaning no branching) or branched. This property of carbon is mainly due to strong carbon-carbon bond and tetra-covalency of carbon.

e. Bonding: Tetra-covalence of Carbon

The theory of bonding was proposed as early as 1916 by Professor Gilbert Lewis. He suggested that atoms can react in such a way as to achieve the electronic configuration of inert gases which had complete valence shell and hence very stable or non-reactive. The reaction between atoms to achieve the stable electronic arrangement results in *chemical bonding*. The formation of bonds can take place in different ways such as (i) *ionic bonds* are formed by transfer of one or more valence electrons from one atom to another. The atom from which the electron is transferred becomes positively charged and the atom receiving the electron becomes negatively charged. For example, sodium chloride is an ionic compound in which sodium loses an electron to chlorine. Thus, sodium becomes positively charged and chlorine becomes negatively charged. Sodium acquires the electronic configuration of neon while the chlorine achieves the electronic configuration of argon; (ii) Atoms which are neither strongly electropositive nor strongly electronegative form bonds by sharing of an electron pair rather than complete transfer of electrons. Such bonding is called *covalent bonding*. Two or more atoms joined by covalent bond(s) form molecules.

Carbon forms four covalent bonds in its compounds. There are four valence electrons in carbon. Since carbon is in the fourth group of periodic table, it is neither highly electronegative nor highly electropositive. It, therefore, does not gain or lose electrons easily to complete its octet. Carbon easily shares its four valence electrons with four other valence electrons of other atoms to form four covalent bonds.

The Lewis dot structures for methane and tetrachloromethane are shown below (**Figure 1.6**). The x represents the valence electrons of carbon and dots represent the valence electrons of hydrogen and chlorine.

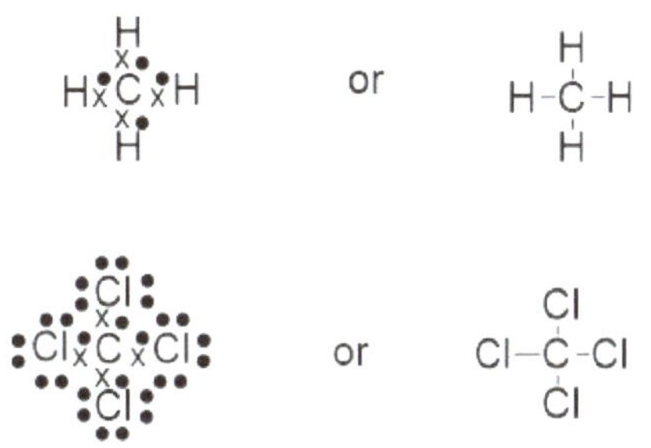

Fig. 1.6: *The Lewis dot structures for methane and tetrachloromethane.*

However, the ground state electronic configuration of carbon is $1s^2$, $2s^2$, $2p_x^{\ 1}$, $2p_y^{\ 1}$, $2p_z^{\ 0}$ (**Figure 1.7**). There are, thus, only two unpaired electrons in the valence shell. Based on this electronic configuration, one would assume the carbon to be bivalent. To form four covalent bonds, one of the 2*s* electrons needs to be promoted to $2p_z$ orbital.

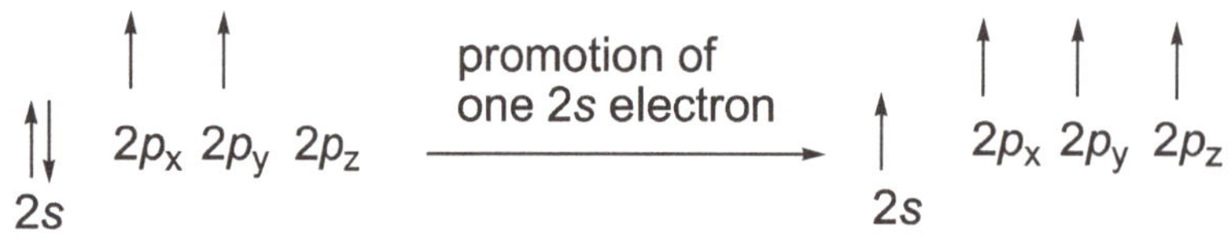

$$\underset{2s}{\uparrow\downarrow}\ \ \underset{2p_x}{\uparrow}\ \underset{2p_y}{\uparrow}\ \underset{2p_z}{}\ \xrightarrow{\text{promotion of one } 2s \text{ electron}}\ \underset{2s}{\uparrow}\ \ \underset{2p_x}{\uparrow}\ \underset{2p_y}{\uparrow}\ \underset{2p_z}{\uparrow}$$

Fig. 1.7: *Electronic configuration of carbon.*

These electron dot structures, however, have limitations in explaining the three-dimensional geometry of the molecules. Furthermore, the energy of 2*s* and 2*p* orbitals are not same. As a result, all the covalent bonds formed would not be identical which was contrary to observations during studies of simple organic compounds. Later, the bonding in carbon compounds was explained based on *molecular orbital theory*. The detail of the bonding in organic compounds based on molecular orbital theory is discussed in the next chapter.

f. Non-Polar and Polar Covalent Bonds

The covalent bonds formed between the identical atoms are non-polar. For example, the covalent bonds between two chlorine atoms (Cl-Cl) or two hydrogen atoms (H-H) are non-polar covalent bonds. The electron pair is shared equally between the two atoms. Covalent bonds, however, can also be formed between two different atoms. If the two atoms involved in formation of a covalent bond are different and have large difference in their electronegativities, e.g., carbon and chlorine, the resulting covalent bond will be polar in nature. Due to large difference in electronegativity, the electron pair is not shared equally. The more electronegative atom pulls electrons towards itself and hence carries a partial negative charge while the less electronegative atom carries a partial positive charge. See the structure of compound called tetrachloromethane below (**Figure 1.8**). In this compound, one carbon is bonded to four chlorine atoms. Since carbon atom is less electronegative than chlorine atom, the carbon atom carries a partial positive charge. The chlorine atoms being more electronegative than carbon atom bear partial negative charge.

$$\begin{array}{ccccc} & & Cl^{\delta\ominus} & & \\ & & | & & \\ Cl^{\delta\ominus} & — & C^{\delta\oplus} & — & Cl^{\delta\ominus} \\ & & | & & \\ & & Cl^{\delta\ominus} & & \end{array}$$

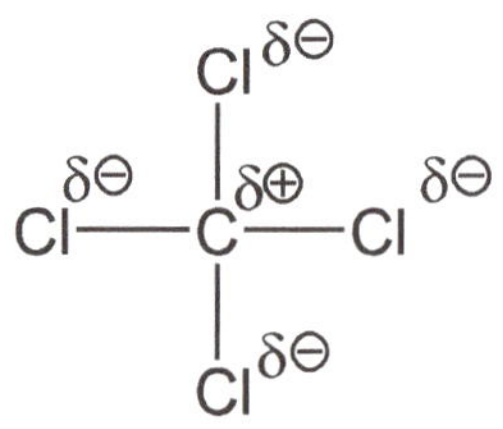

Fig. 1.8: *Structure of tetrachloromethane showing polarity of C-Cl bonds.*

SUMMARY

Organic chemistry is a very important branch of chemical sciences. It deals with the chemistry of carbon compounds. Urea is the first organic compound synthesized. Organic compounds are extremely useful in day-to-day life. Most medicines and household items like clothing, furniture, floor tiles, and soaps are derived from organic compounds. For a systematic study of millions of organic compounds, they are divided into certain groups depending on the functional group present in them. A functional group in an organic compound is responsible for the property of that compound. All compounds with the same functional group possess same chemical properties. The compounds with a particular functional group can be further arranged in a homologous series depending on the molecular weight. The compounds of a homologous series show a trend in their physical properties such as melting point, boiling point, and solubility. The carbon atom in organic compounds shows tetracovalence that is the carbon atom will have four covalent bonds. The covalent bond may be non-polar or polar depending on the electronegativity difference between the carbon and the other atom attached.

EXERCISES

1. Identify the following compounds as organic or inorganic compounds.

   a. Sodium carbonate (Na_2CO_3),

   b. methane (CH_4)

   c. Spirit (C_2H_5OH)

   d. calcium carbide (CaC_2)

   e. sodium nitrite ($NaNO_2$)

   f. nitromethane (CH_3NO_2)

   g. acetic acid (CH_3COOH)

   h. calcium sulfate ($CaSO_4$)

2. Who synthesized the first organic compound in the laboratory? What was the name of this first organic compound?

3. Name at least five organic compounds which you know from your day-to-day life experience.

4. Select hydrocarbons from the list of the following compounds.

   a. CH_3CH_3, b. CH_3NH_2, c. NH_3, d. C_2H_5OH, e. C_6H_6, f. C_2H_4, g. $C_{10}H_8$.

5. How many electrons are there in the valence shell of carbon atom?

6. How many unpaired electrons are there in the valence shell of carbon atom?

7. Based on your answers of the above two exercises, how many covalent bonds do you expect from a carbon atom to form.

8. What valance the carbon atom shows in organic compounds?

9. Which of the following compounds belong to the same homologous series?

   a. CH_3CH_2OH, b. CH_3OCH_3, c. CH_3OH, d. CH_3COOH, e. $(CH_3)_2CHOH$,
   f. $CH_3CH_2CH_2CH_2CH_2CH_2OH$, g. C_6H_5OH

10. Refer to the functional groups under Section 1.5 and mention the class of the following organic compounds.

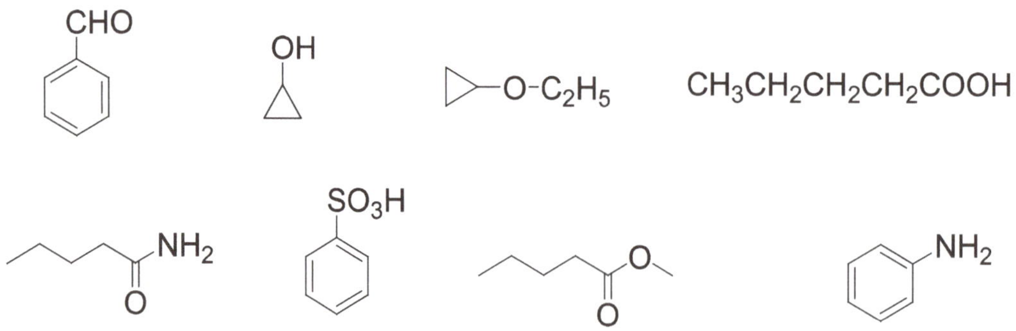

CHO   OH $-O-C_2H_5$ $CH_3CH_2CH_2CH_2COOH$

NH_2 / O SO_3H O / O NH_2

11. The structural formulas shown below have the same molecular formula C_4H_8. What is the relationship between the two structures?

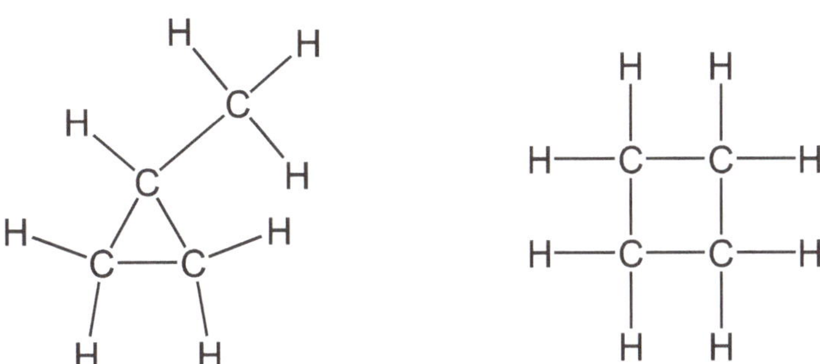

12. The following pairs of compounds have the same atom connectivity in their corresponding structural formula. What kind of isomers are they?

(a) H_3C, CH_3 / H, H and H, CH_3 / H_3C, H (b) Br, Br / H, H and H, Br / Br, H

13. Draw the structure of chloroform (CBr_4, carbon is the central atom) and indicate bond polarity using δ+/δ- convention.

14. Select the most and least polar covalent bonds in compounds shown below. Give reason for your answer.

H_3C-OH H_3C-F H_3C-NH_2

BONDING, STRUCTURE, REACTIONS AND MECHANISMS, ACIDS AND BASES

As mentioned in the previous chapter, for a carbon atom to form four covalent bonds, an electron from 2*s* orbital gets promoted to 2*p* orbital resulting into four half-filled orbitals in the valence shell. Each of these half-filled orbitals can now accommodate an additional electron to become fully filled and form four covalent bonds (**Figure 1.7**). We also need to account for finding that the all four covalent bonds of methane are identical in terms of bond length and bond energy which cannot be explained just by promotion of one 2*s* electron to 2*p* orbital because the *s* and *p*-orbitals do not have the same energy level. The bond formed from *s*-orbital electron would have different energy and length from the bonds formed from *p*-orbital electrons. This is to be explained in this chapter using the molecular orbital theory and the concept of hybridization. However, before going into further details of the bonding in organic compounds based on molecular orbital theory, it is important to recall the shape of orbitals and the way they overlap to form sigma (σ) and pi (π) covalent bonds.

2.1 Shape of *s*- and *p*-Orbitals and Formation of σ- and π-Bonds

The shapes of *s* and *p* orbitals are different (**Figure 2.1**). The s orbitals have spherical shape. The three *p* orbitals are dumbbell shaped and mutually perpendicular, oriented along the three-coordinate axis, x, y, and z.

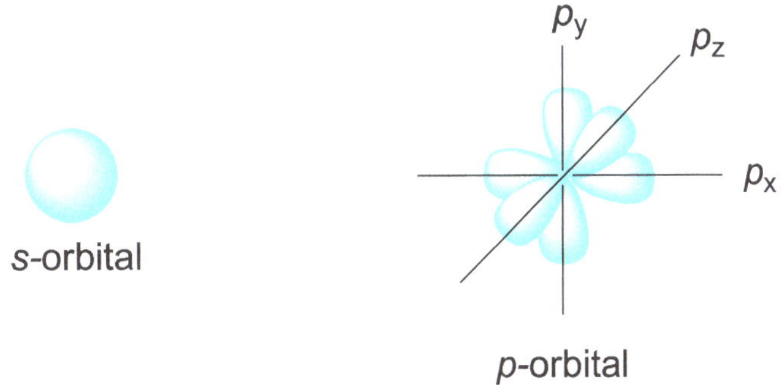

Fig. 2.1: *The shapes of s and p orbitals.*

In the orbital approach of bonding, the two atomic orbitals containing valence electrons overlap with each other to form **molecular orbitals**. Like atomic orbitals, each molecular orbital can accommodate a maximum of two electrons. Let's take the example of hydrogen molecule formation (**Figure 2.2**). The 1*s* atomic orbitals of two hydrogen atoms overlap to form the H-H molecular orbital or a single covalent bond known as sigma (σ) bond.

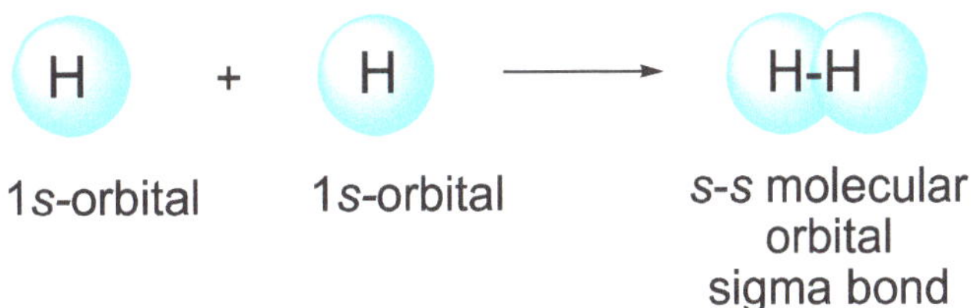

Fig. 2.2: *Formation of hydrogen molecule.*

The overlapping of an *s*-orbital with a *p*-orbital or overlapping of two *p*-orbitals in head-on fashion also forms covalent bond called sigma (σ) bond (**Figure 2.3**). The covalent bond formed by lateral (sidewise) overlap of two *p*-orbitals is called a pi (π) bond.

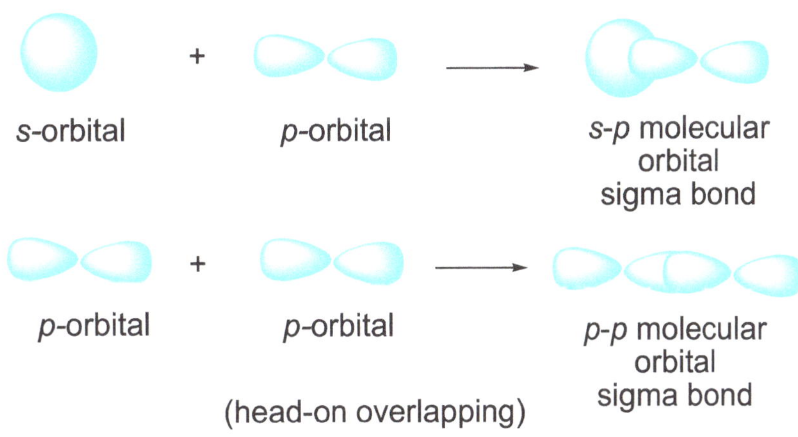

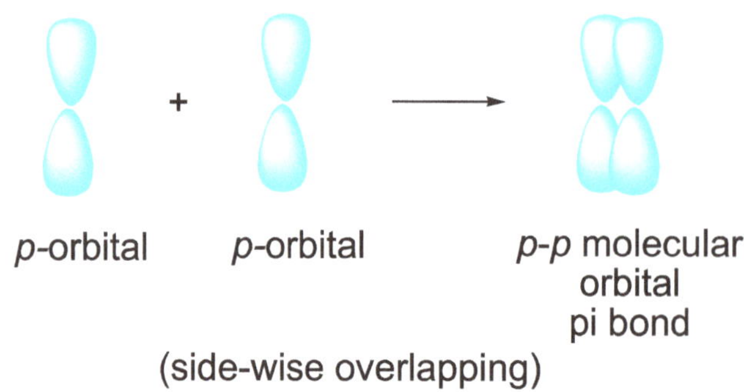

Fig. 2.3: *Formation of sigma- and pi-bonds.*

a. Types of Bonds

Now you can generalize the above discussion as follows.

a. All single bonds would be sigma bonds.

b. A double bond would be made up of a sigma and a pi bond.

c. A triple bond would be made up of a sigma and two pi bonds.

Of the two types of covalent bond, σ and π, the π-bond is weaker because the electrons are loosely bound in lateral (sidewise) overlap of the orbitals.

b. Bond lengths and Bond strength

The distance between the nuclei of two atoms in a bond is called bond length. It depends on the atoms involved in bonding and the number of bonding electron pairs. The greater the number of electron pairs involved in bonding shorter would be the bond length. This is because the electron pairs pull the nuclei towards themselves reducing the distance between the two nuclei. Thus, carbon-carbon triple bond will be the shortest and carbon-carbon single bond will be the longest.

$$\underset{\text{Longest}}{C{-}C} > C{=}C > \underset{\text{Shortest}}{C{\equiv}C}$$

Bond strength is the energy required to break the bond. More energy will be required to break a strong bond and less energy will be required to break a weak bond. The bond strength increases as the bond length decreases (**Table 2.1**). Thus, the least energy is required to break a carbon-carbon single bond and highest energy is required to break the carbon-carbon triple bond.

C—C $<$ C=C $<$ C≡C

Weakest Strongest

Table 2.1: *Bond length and bond strength of carbon-carbon bonds.*

Bond type	Bond length	Bond strength
C—C	1.54 Å	348 kJ/mol
C=C	1.33 Å	614 kJ/mol
C≡C	1.20 Å	839 kJ/mol

2.2 Bonding in Organic Compounds: Molecular Orbital Approach

The bonding in organic compounds is quite complex in comparison to straightforward-looking hydrogen molecule. The simplest organic compound methane has four covalent bonds (four carbon-hydrogen single bonds). The simplest alkene, ethylene (C_2H_4) has six covalent bonds in total. Each of the two carbon atoms forms two covalent bonds with two hydrogen atoms (thus a total of four carbon-hydrogen single bonds). Two covalent bonds are between two carbon atoms (one carbon-carbon double bond). The simplest alkyne, acetylene (C_2H_2) has five covalent bonds. Each of the two carbon atoms forms one covalent bond with one hydrogen atom (thus a total of two carbon-hydrogen single bonds). The two carbon atoms in acetylene are attached by three covalent bonds (one carbon-carbon triple bond).

It has been mentioned in the beginning that one electron from the 2*s* orbital of the carbon atom is promoted to the 2*p* orbital so that the valence shell of carbon could have four unpaired electrons to share with other atoms. Now you can assume that, in methane, carbon shares four of its unpaired electrons with four electrons, one each from four hydrogen atoms, to form four covalent bonds. The problem of bonding in carbon compounds, however, is not completely resolved by this explanation.

The study of methane, CH_4, showed that all four covalent bonds in it were identical having the same length, energy, and bond angle even though the bond forming orbitals 2s and 2p had different energy levels. The four covalent bonds were directed towards the corner of a regular tetrahedron. The monochlorination of methane (substitution of one hydrogen atom of methane with a chlorine atom) gives only one chloromethane. If all the carbon-carbon bonds would not be equivalent, there would be formation of isomeric chloromethane molecules. An answer to this problem was provided in 1931 by Linus Pauling using the concept of hybridization.

Hybridization

Hybridization is the mixing of atomic orbitals of nearly equivalent energy and redistribution of energy among themselves leading to the formation of equal number of hybrid orbitals (**Figure 2.4**) having same energy. Thus, the 2s and 2p orbitals mix in excited state of carbon atom and redistribute the energy among themselves to form equal number of hybrid orbitals with same energy.

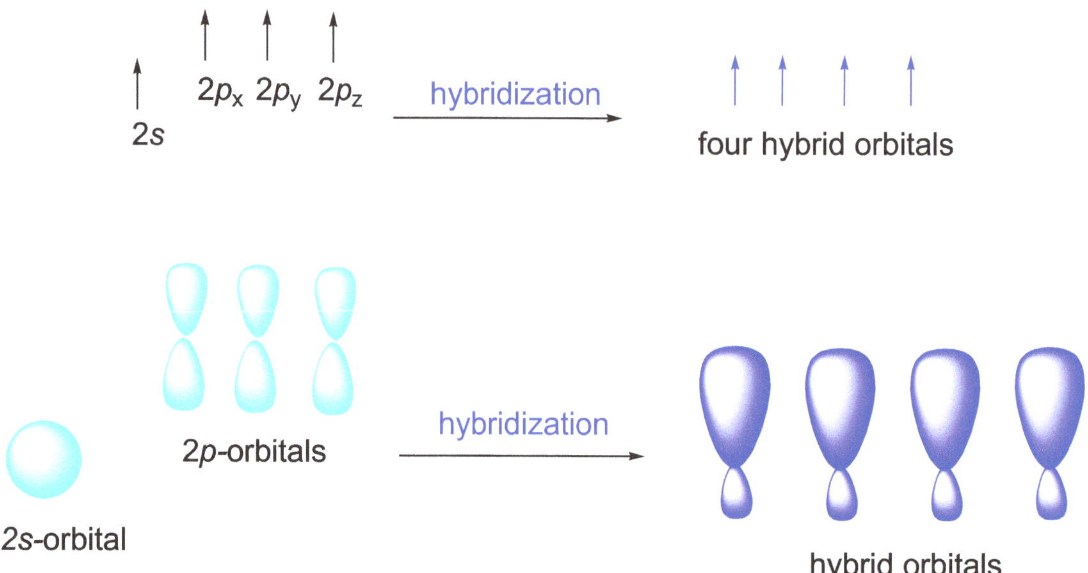

Fig. 2.4: *Formation of hybrid orbitals of carbon.*

Hybridization in carbon can occur in three possible ways.

i. sp³ Hybridization: Structure of Methane

One 2*s* orbital and three 2*p* orbitals hybridize to form four hybrid orbitals, each called sp^3 hybrid orbital, all having same energy (**Figure 2.5**). They arrange themselves equally in space leading to a tetrahedral shape. These hybridized orbitals are used to form σ-bonds (sigma bonds) with four hydrogen atoms. The bond angle between any of the two bonds formed is 109.5° (**Figure 2.6**).

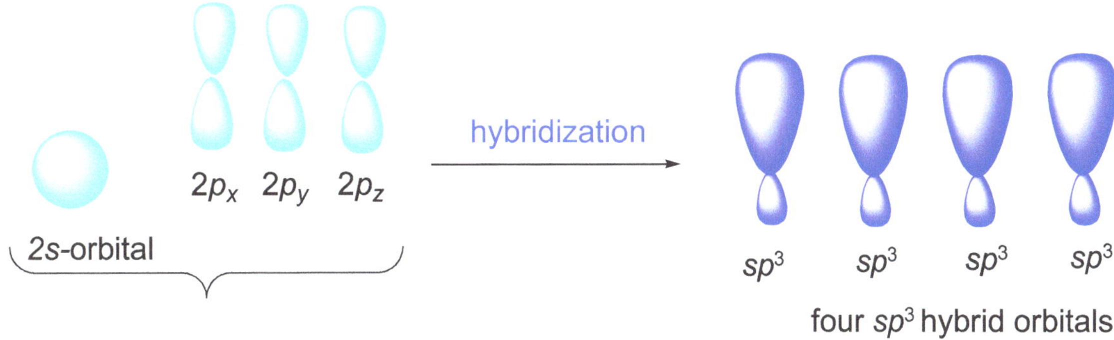

Fig. 2.5: *sp^3 Hybridization in carbon.*

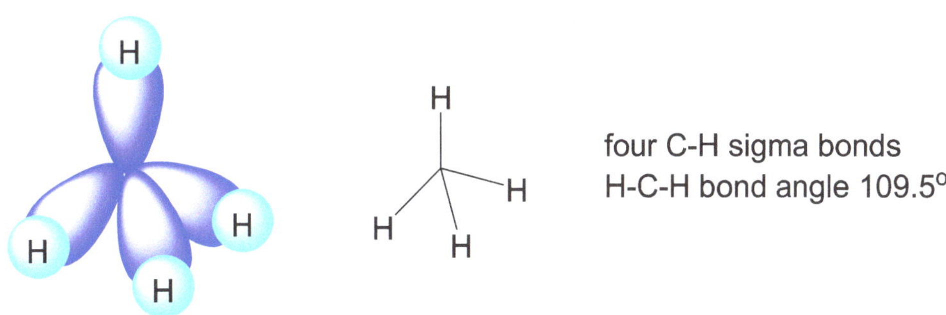

Fig. 2.6: *Molecular orbital picture of methane.*

In this way, ethane ($H_3C\text{-}CH_3$) will have seven σ-bonds. Each carbon has three σ-bonds with three hydrogen atoms and one σ-bond between the two carbon atoms.

ii. sp^2 Hybridization: Structure of Ethylene

In ethylene ($H_2C{=}CH_2$), one 2*s* orbital and two 2*p* orbitals of the carbon atom hybridize to form three equivalent hybrid orbitals called sp^2 hybrid orbitals (**Figure 2.7**).

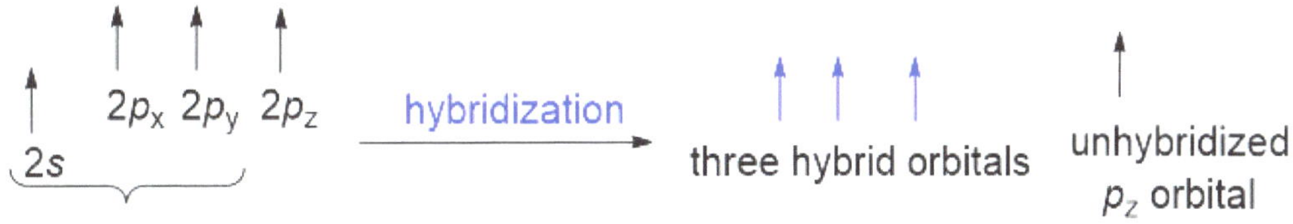

Fig. 2.7: *sp^2 Hybridization in carbon.*

The three sp^2 hybrid orbitals arrange themselves in a trigonal planar shape in space and form σ-bonds with other atoms. The angle between any two bonds is 120°. In ethylene, there are two carbon atoms, so there will be six sp^2 hybrid orbitals in all. Four sp^2 hybrid orbitals, two from each carbon, are used to form covalent bonds with *s* orbitals of four hydrogen atoms. The remaining two sp^2 hybrid orbitals are used to form a covalent bond between two carbon atoms. Remember that all these five covalent bonds are σ-bonds.

The remaining one 2*p* orbital on each carbon remains unhybridized. It exists perpendicular to the plane of hybrid orbitals. The unhybridized *p*-orbitals on carbon atoms overlap sidewise (laterally) leading to the formation of a covalent bond that is a π-bond. Thus, in the structure shown below (**Figure 2.8**), one of the two bonds between two carbon atoms is a σ-bond while the other is a π-bond. Since the π-bond is loosely bound due to sidewise overlapping, the electron-pair involved in a π-bond formation is readily available for reaction with an electron-deficient species. Therefore, hydrocarbons containing double bonds or triple bonds are generally more reactive in comparison to saturated hydrocarbons.

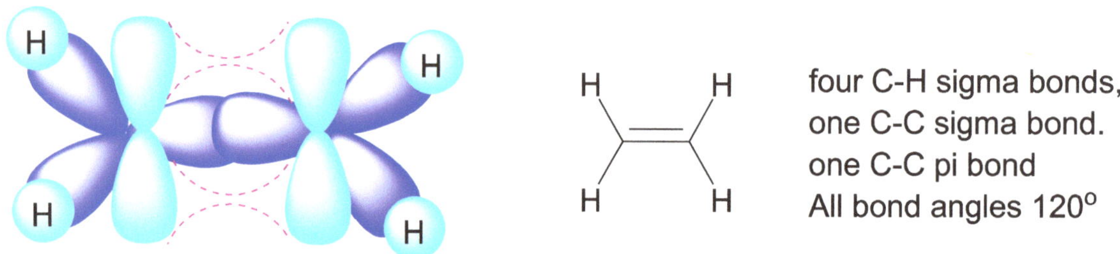

Fig. 2.8: *Molecular orbital picture of ethene.*

iii. sp Hybridization: Structure of Ethyne (Acetylene)

One 2*s* orbital and only one of the 2*p* orbitals hybridize resulting in the formation of two equivalent hybrid orbitals called *sp* hybrid orbitals (**Figure 2.9**). These orbitals extend in opposite directions from the center taking a linear geometry. They form two σ bonds with bond angles of 180°. The remaining two 2*p* orbitals, which are perpendicular to each other and perpendicular to the hybrid orbitals as well, overlap laterally with properly aligned (parallel) orbitals on adjacent atoms to form two pi bonds. In acetylene, there are two carbon atoms (**Figure 2.10**). Each carbon atom forms a sigma bond with a hydrogen atom. The remaining *sp*-hybridized orbital – one on each carbon is used to form a sigma bond between two carbon atoms. In total, there are three sigma bonds in acetylene. Of the three covalent bonds between the two carbon atoms, two are pi bonds.

Fig. 2.9: *sp Hybridization in carbon.*

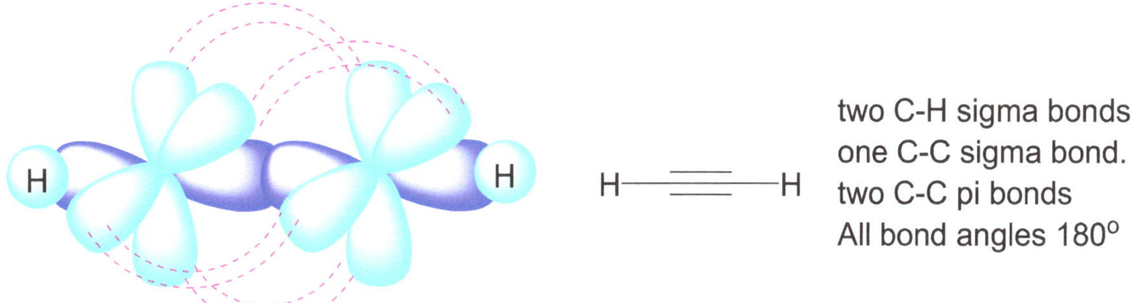

Fig. 2.10: *Molecular orbital picture of ethyne.*

In summary, the geometry and hybridization of carbon atom in organic compounds can be described as follows.

a. A carbon attached to four single bonds - sp^3-hybridized, tetrahedral geometry, bond angle of 109.5°.

b. A carbon with a double bond - sp^2-hybridized, trigonal planar geometry, bond angle of 120°.

c. A carbon with two double bonds (=C=) - *sp*-hybridized, (=C=); linear geometry, bond angle of 180°.

d. A carbon with a triple bond - *sp*-hybridized, linear geometry, bond angle of 180°.

Problem: Identify the hybridization of carbon atoms in the carbon-carbon bonds indicated with arrows in the following molecule.

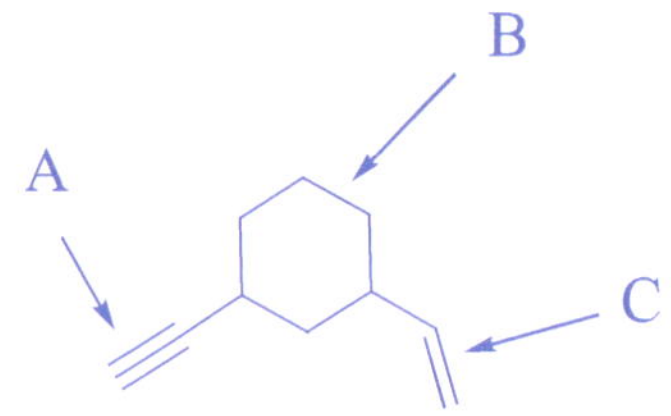

A B C

Solution: A. *sp* B. sp^3 C. sp^2

2.3 Structural Formulas of Organic Compounds

To study the chemical properties of organic compounds the knowledge of molecular formula alone is not sufficient. One needs to know the structural formula or simply structure of the compounds. There are various ways of representing a structure. For example, a complete or detailed structural formula shows all the atoms and bonds between them (**Figure 2.11**).

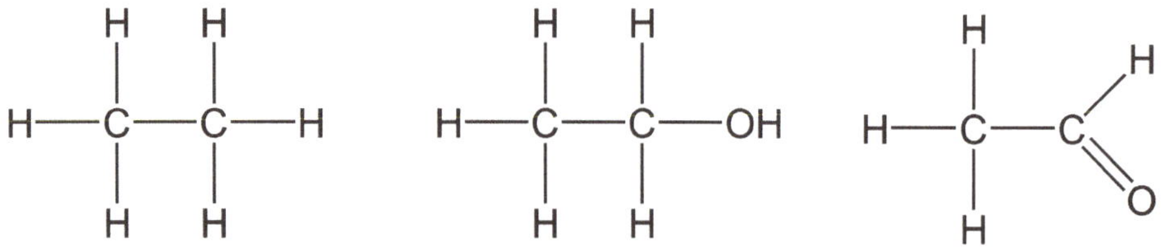

Fig. 2.11: *Complete structural formulas.*

It is, however, not always necessary to draw a complete structural formula; instead, either a **condensed or an abbreviated structural formula** is drawn (**Figure 2.12**). This shows the groups of atoms and bonds between carbon atoms only, though the latter is not essential.

CH_3-CH_2-CH_3
OR
$CH_3CH_2CH_3$

O═$CHCH_2CH_3$ HO—$CH_2CH_2CH_3$

Fig. 2.12: *Condensed structural formulas.*

Alternatively, the **line formulas**, which show only the bond between two carbon atoms or between the atoms heavier than a hydrogen atom (**Figure 2.13**), are used. Even the carbon is not shown; it is understood to be on every corner and at every end.

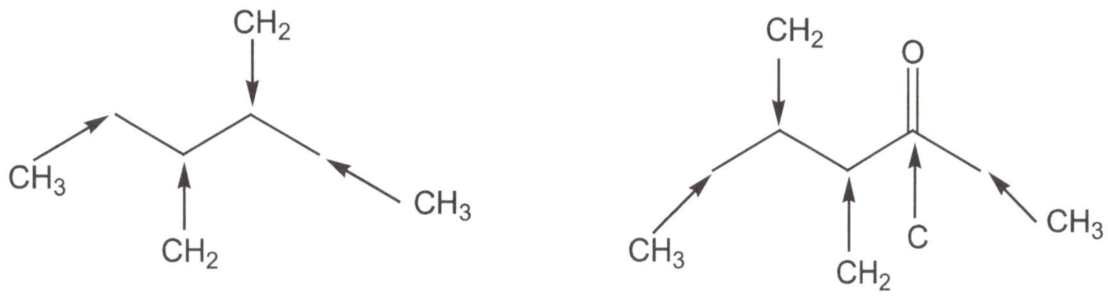
CH2
CH3
CH2
CH3
CH2
O
CH3
C
CH3
CH2

Fig. 2.13: *Line formulas.*

In practice, however, it is not always necessary to write a pure condensed or pure line formula. Many times, only that part of the structure is shown where a chemical reaction is to be explained.

2.4 General Reactions of Organic Compounds

Organic compounds undergo numerous types of chemical reactions. These reactions can be broadly classified into the following four types.

a. Substitution

One or more atoms in a compound are substituted by other atom(s). Substitution reactions are common in saturated compounds, e.g., alkanes, haloalkanes. Aromatic hydrocarbons (arenes) also undergo substitution reactions under the influence of certain catalysts.

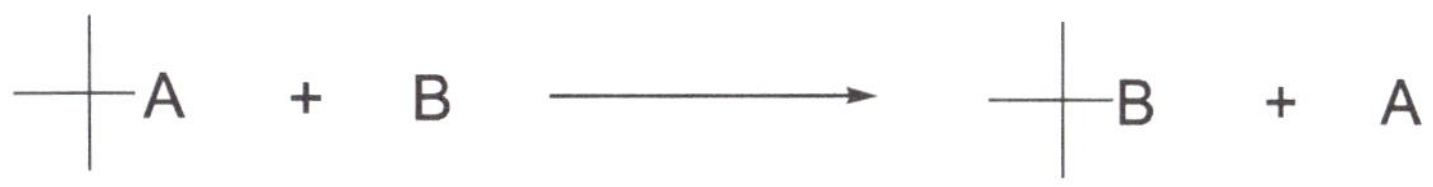
A
+
B
B
+
A

For example:

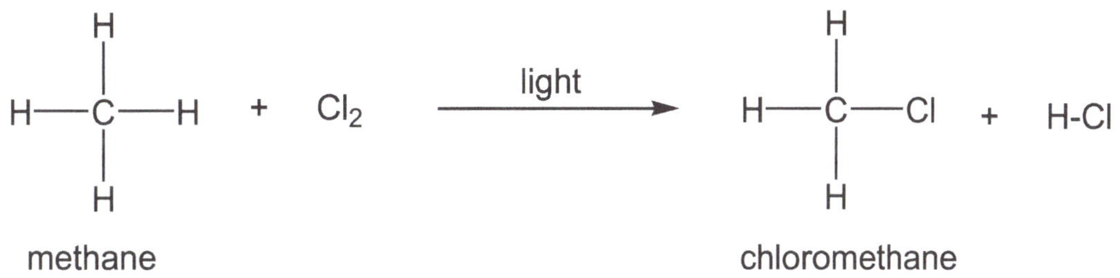
H
H
C
H
H
+
Cl2
light
H
H
C
Cl
H
+
H-Cl
methane
chloromethane

b. Addition

Addition reactions are those in which the reagent is completely added to the organic compound. It is common in unsaturated compounds (compounds with multiple bonds) such as alkenes, alkynes, and carbonyl compounds.

$$>C=C< \; + \; A\text{-}B \longrightarrow -\underset{A}{\underset{|}{\overset{|}{C}}}-\underset{B}{\underset{|}{\overset{|}{C}}}-$$

For example:

$$H_2C=CH_2 \; + \; H\text{-}Cl \longrightarrow H-\underset{H}{\underset{|}{\overset{\overset{H}{|}}{C}}}-\underset{H}{\underset{|}{\overset{\overset{H}{|}}{C}}}-Cl$$

ethylene — chloroethane

c. Elimination

Elimination reactions can be treated as the reverse of an addition reaction. Organic molecule fragments in such a way as to give another organic compound with an unsaturated center.

$$-\underset{A}{\underset{|}{\overset{|}{C}}}-\underset{B}{\underset{|}{\overset{|}{C}}}- \longrightarrow >C=C< \; + \; A\text{-}B$$

For example:

$$H-\underset{H}{\underset{|}{\overset{\overset{H}{|}}{C}}}-\underset{CH_3}{\underset{|}{\overset{\overset{CH_3}{|}}{C}}}-OH \xrightarrow{H^+} H_2C=C(CH_3)_2 \; + \; H_2O$$

d. Oxidation and Reduction

Several classes of organic compounds can be oxidized or reduced by reacting with appropriate oxidizing or reducing agents. Oxidation of hydrocarbons is commonly referred to as combustion. Hydrocarbons get oxidized by oxygen forming carbon dioxide and water. The reduction of multiple bonds by addition of hydrogen atoms across the multiple bonds is referred as hydrogenation as well. Examples of oxidation of an alcohol, ethanol to acetaldehyde and reduction of the latter to ethanol are shown below.

$$CH_3CH_2OH \xrightarrow{\text{oxidation}} CH_3CHO$$

ethanol → acetaldehyde

$$CH_3CHO \xrightarrow{\text{Reduction}} CH_3CH_2OH$$

acetaldehyde → ethanol

e. Molecular Rearrangements

Molecular rearrangements involve migration of an atom or group from one atom to another within the molecule.

$$C\text{—}A\text{—}B \longrightarrow A\text{—}B\text{—}C$$

2.5 Reaction Mechanisms: Cleavage of Covalent Bond

The chemical equation for any reaction tells us about the reactants taking part in the chemical reaction, reaction conditions, and products formed in it. A chemical reaction involves breaking of bond in the reactants and formation of new bonds. As a result of bond-breaking unstable species are formed in the reaction. These unstable species are called Reactive Intermediates. These intermediates, because they are unstable, quickly combine further in different fashions giving rise to stable products.

A reaction mechanism is a step-by-step description of the bond breaking and bond-forming in a chemical reaction. The reaction mechanism of a reaction is determined by several other studies. Before knowing how to show this bond breaking and forming schematically, let's have an idea of two types of breaking of bonds.

A covalent bond can break either by a homolytic or by a heterolytic fission. In the former type, the electrons of the covalent bond are equally shared by departing atoms. The intermediates formed are called free radicals or simply radicals. In heterolytic fission, both electrons go with one atom. The atom taking electrons will be negatively charged whereas the other atom losing its share of electron will be positively charged. An intermediate with negative charge on carbon is called carbanion whereas the one with positive charge on carbon is called carbocation. Those reactions that proceed through radicals are said to follow the radical mechanism. Similarly, those reactions involving ionic intermediates are called to follow the ionic mechanism. There are several physico-chemical methods that are used to determine whether a reaction is proceeding through radicals or ions.

$$-\overset{|}{\underset{|}{C}}:L \xrightarrow{\text{heterolytic cleavage}} -\overset{|}{\underset{|}{C}}^{\oplus} + L:$$

A carbocation

$$-\overset{|}{\underset{|}{C}}:L \xrightarrow{\text{heterolytic cleavage}} -\overset{|}{\underset{|}{C}}:^{\ominus} + L$$

A carbanion

$$-\overset{|}{\underset{|}{C}}:L \xrightarrow{\text{homolytic cleavage}} -\overset{|}{\underset{|}{C}}\bullet + L\bullet$$

A free radical

2.6 Arrow Formalism in Organic Reactions

In writing the reaction mechanisms, different types of arrows are used. They have specific meanings.

a. Straight Arrows

They might be single, double, and half-headed. A single headed straight arrow points from reactant to the product side. A pair of half-headed straight arrows heading in opposite directions is commonly used to indicate that the reaction is reversible. A double-headed straight arrow is used between two resonance structures. It does not indicate occurrence of any chemical reaction, just the different arrangement of pi electrons in the same molecule.

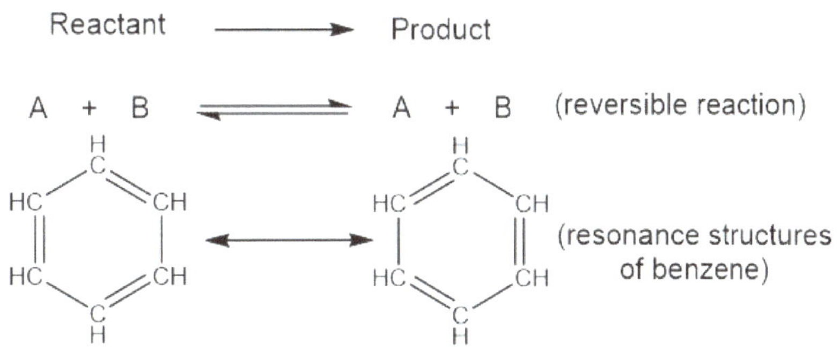

Reactant $\longrightarrow$ Product

A + B $\rightleftharpoons$ A + B (reversible reaction)

(resonance structures of benzene)

b. Curved Arrows

The curved arrows are used to show the movement of electrons in the molecule. A single-headed curved arrow always originates from an electron-rich center and ends at an electron-deficient center or at more electronegative atom. In the first equation shown below, the π electrons pick a proton and form a covalent bond (C-H). In the second equation, the π electrons move to more electronegative oxygen atom. Hence, the oxygen atom bears a negative charge and carbon a positive charge. A half-headed curved arrow, called a fishhook is used to indicate the movement of a single electron. Such arrows are shown in the third equation below to show the hemolytic cleavage of the carbon-carbon single bond forming two methyl radicals.

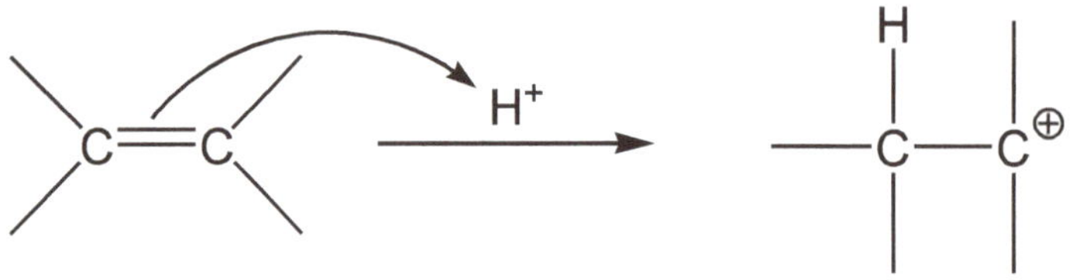

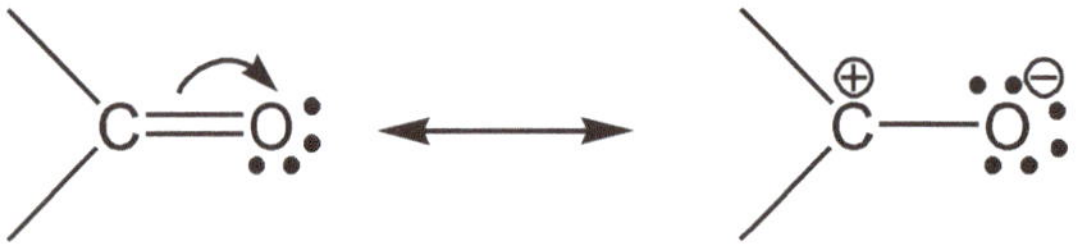

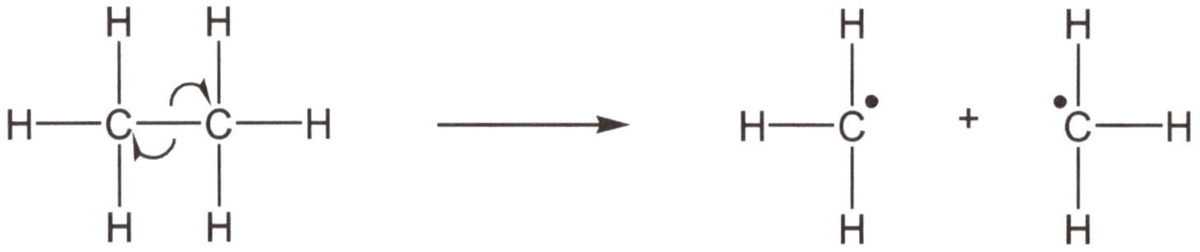

2.7 Acids and Bases

a. Brønsted – Lowry Acids and Bases

The Danish scientist Johannes Brønsted and the British scientist Thomas Lowry, in 1923, defined acid as any substance, that can donate a proton to another substance. Such substances are now called Brønsted-Lowry acids and bases after the name of the discoverers. This substance might be a neutral molecule or an ion. See examples of acetic acid, HSO_4^-, and HCO_3^- in equations below that donate a proton to water changing it to H_3O^+ called hydronium ion.

CH_3COOH (aq) + H_2O (l) H_3O^+ (aq) + CH_3COO^- (aq)

HSO_4^- (aq) + H_2O (l) H_3O^+ (aq) + SO_4^{2-} (aq)

HCO_3^- (aq) + H_2O (l) H_3O^+ (aq) + CO_3^{2-} (aq)

A base is a substance including ions, which can accept a proton from another substance. For example,

HO^- (aq) + H^+ (aq) H_2O (l)

NH_3 (aq) + H_2O (l) NH_4^+ (aq) + OH^- (aq)

R-NH_2 (aq) + H_2O (l) R-NH_3^+ (aq) + OH^- (aq)

CH_3COO^- (aq) + H_2O (l) CH_3COOH (aq) + OH^- (aq)

b. Strong and Weak Acids and Bases

Those acids, which ionize completely in water, are called strong acids. The strong acids include hydrochloric acid, sulfuric acid, and nitric acid etc. A weak acid does not ionize completely in water. Organic acids such as acetic acid (ethanoic acid), propanoic acid, etc. fall in this category. Similarly, a strong base reacts completely with water to produce hydroxide ion, that is, it is fully protonated in water. Metal hydroxides such as sodium hydroxide, and potassium hydroxide are strong bases. A weak base reacts incompletely with water and thus does not get fully protonated. Ammonia and amines are weak bases. In case of weak acids and bases, equilibrium is established before the completion of reaction.

$$CH_3COOH\ (aq) + H_2O\ (l) \rightleftharpoons H_3O^+\ (aq) + CH_3COO^-\ (aq)$$

$$NH_3\ (aq) + H_2O\ (l) \rightleftharpoons NH_4^+\ (aq) + HO^-\ (aq)$$

The strength of an acid in water solution can be expressed by its acidity constant K_a. For a general acid HA,

$$K_a = \frac{[H_3O^-]\ [OH^-]}{[HA]}$$

The stronger the acid, the more this equilibrium is shifted to the right. Thus, increasing the concentration of hydronium ion H_3O^+ increases the value of K_a.

For water, which is an amphiprotic liquid, this expression and the value of K_a is shown below.

$$K_a = \frac{[H_3O^-]\ [OH^-]}{[H_2O]} = 1.8 \times 10^{-14}$$

The acidity is often expressed as pK_a which is negative logarithm of acidity constant K_a. This avoids the use of negative values as shown in expression for ionization of water.

pK_a = - log K_a

Thus, pK_a decreases with increasing K_a. This means that the stronger the acid, the lower will be its pK_a and weaker the acid, higher will be its pK_a. Furthermore, an inverse relationship exists between the strength of an acid and its conjugate base. Thus, the conjugate base of a strong acid will be weaker, and the conjugate base of a weak acid will be stronger. This is known as conjugate seesaw relationship.

$$HCl\,(aq) + H_2O\,(l) \rightleftharpoons H_3O^+\,(aq) + Cl^-\,(aq)$$

strong acid ← seesaw relationship → weak conjugate base

$$CH_3COOH\,(aq) + H_2O\,(l) \rightleftharpoons CH_3COO^-\,(aq) + H_3O^+\,(aq)$$

weak acid ← seesaw relationship → weak conjugate base

c. Lewis Acids and Bases

The Lewis acid is any substance that has a vacant orbital and can thus accept an electron pair. The base is any substance that can donate an electron pair. This definition of acids and bases is thus not limited only to the substances that can accept or donate proton. In an acid-base reaction, donation and acceptance of an electron pair leads to the formation of a new covalent bond.

According to this definition, a proton can be defined as an acid because it can accept an electron pair from a base to fill its vacant 1*s* orbital. Now let's see aluminum chloride. It does not have any proton to donate. But it can accept an electron pair and hence it can be defined as an acid according to the Lewis definition. Chloride ion or trimethylamine have ability to donate an electron pair, hence they are Lewis bases.

$$AlCl_3 + :Cl^- \rightleftharpoons AlCl_4^-$$

Lewis acid   Lewis base

$$AlCl_3 + :NH_3 \rightleftharpoons Cl_3Al^{-}-N^{+}H_3$$

Lewis acid   Lewis base

SUMMARY

The molecular orbitals are formed by overlapping of atomic orbitals. The *s*-atomic orbital has a spherical shape whereas the *p*-atomic orbital has a dumbbell shape. The overlapping of two *s*-orbitals or overlapping of a *s*-orbital and a *p*-orbital forms a σ-bond. The head-on overlapping of two *p*-orbitals forms a σ-bond as well while the sidewise overlapping of two *p*-orbitals forms a π-bond. In carbon compounds, one of the 2*s* electrons is promoted to 2*p* orbital resulting in formation of four half-filled atomic orbitals. These orbitals hybridize and form an equal number of hybrid molecular orbitals with same energy. A carbon-carbon single bond (σ-bond) is formed by sp^3-hybridized carbon atoms having tetrahedral geometry (bond angle of 109.5°). A carbon-carbon double bond (one σ-bond and one π-bond) is formed by sp^2 hybridization (trigonal planar geometry, bond angle of 120°) whereas a carbon-carbon triple bond (one σ-bond and two π-bonds) is formed by *sp* hybridization (linear geometry, bond angle of 180°).

The structural formula of organic compounds is drawn in different ways. The detailed structural formula shows all the atoms and bonds between them. To understand the chemistry of a particular compound, however, it is not always necessary to draw a detailed structural formula; one can draw a condensed or line formula.

The reactions of organic compounds can be broadly classified as substitution, addition, elimination, oxidation, reduction, and molecular rearrangement reactions. We write chemical equations for a chemical reaction showing all the substrates, products, and reaction conditions. A detailed study of how the product is formed in a chemical reaction and its stepwise description is called reaction-mechanism. The mechanism of a chemical reaction is confirmed by several other studies. The reactions may occur by an ionic mechanism or free-radical mechanism. Ions are formed by a heterolytic cleavage of covalent bond whereas free radicals are formed by a hemolytic cleavage of covalent bond. The curved arrows are used to show the movement of electrons in a reaction.

Any substance that can donate a proton is defined as an acid whereas any substance that can accept a proton is defined as a base as per Brønsted definition. According to Lewis definition, acids are electron pair acceptors and bases are electron pair donors. A strong acid ionizes completely in solution whereas a weak acid ionizes partially in solution.

EXERCISES

1. How can you explain the formation of four covalent bonds in organic compounds when there are only two unpaired electrons in the valance shell of carbon?

2. All four covalent bonds in methane are identical in terms of bond length and bond energy. How can you explain this observation when the energies of 2*s* and 2*p* orbitals of the carbon atom are different?

3. Briefly explain the two types of overlapping of atomic orbitals that lead to the formation of σ and π bonds. Which of the two bonds σ or π is formed when a 1s atomic orbital of hydrogen atom overlaps with *sp*2-hybridized orbital of carbon atom?

4. Identify the hybridization of each carbon atom in the following compounds. Mention their bond angle and geometry.

5. Count the total number of σ and π bonds in the compound shown in exercise 4.

6. Write the detailed structural formula and molecular formula corresponding to following line formulas.

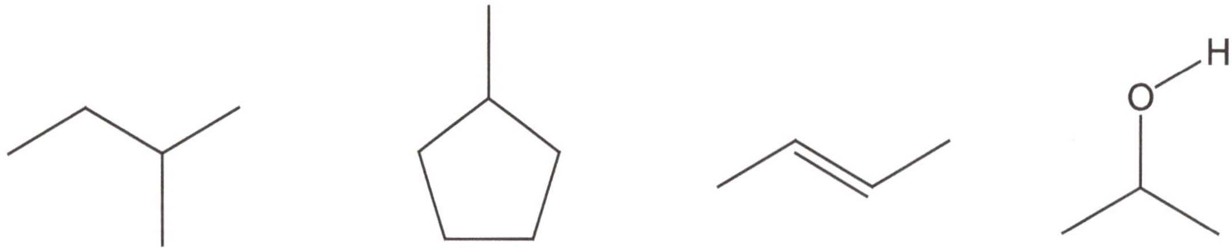

7. Write all possible structural formulas (detailed and line) corresponding to molecular formula C_5H_{12}.

8. Identify the following reactions as substitution, addition, elimination, and oxidation or reduction.

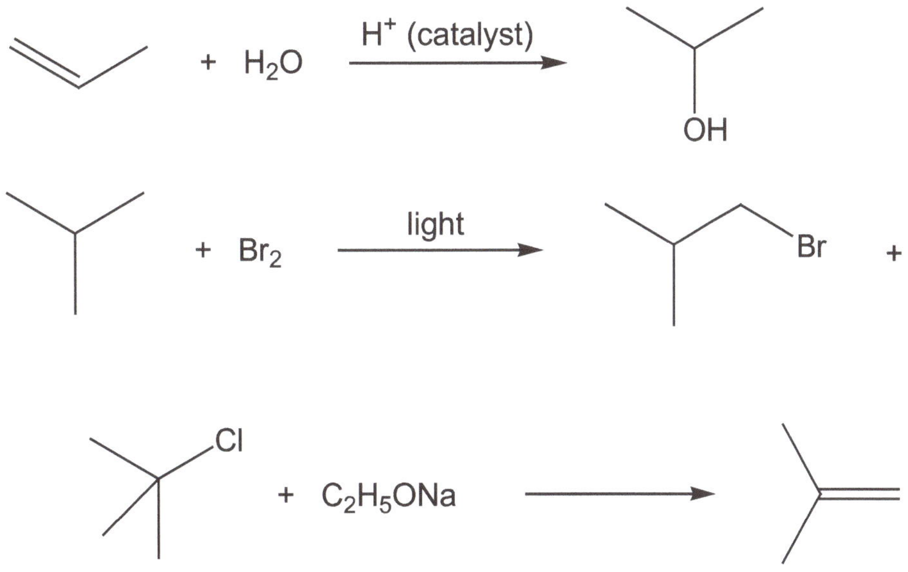

+ H_2O —— H^+ (catalyst) ——→ OH

+ Br_2 —— light ——→ Br + HBr

Cl + C_2H_5ONa ——→

C_3H_8 + 5 O_2 ——→ 3 CO_2 + 4 H_2O

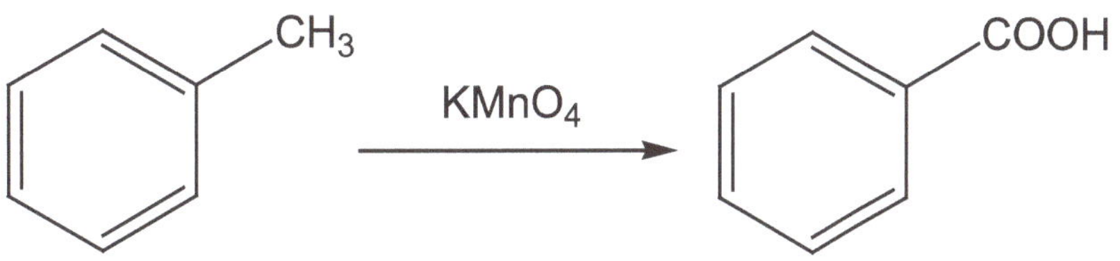

CH_3 —— $KMnO_4$ ——→ COOH

9. What do the curved arrows indicate in the following reaction equations?

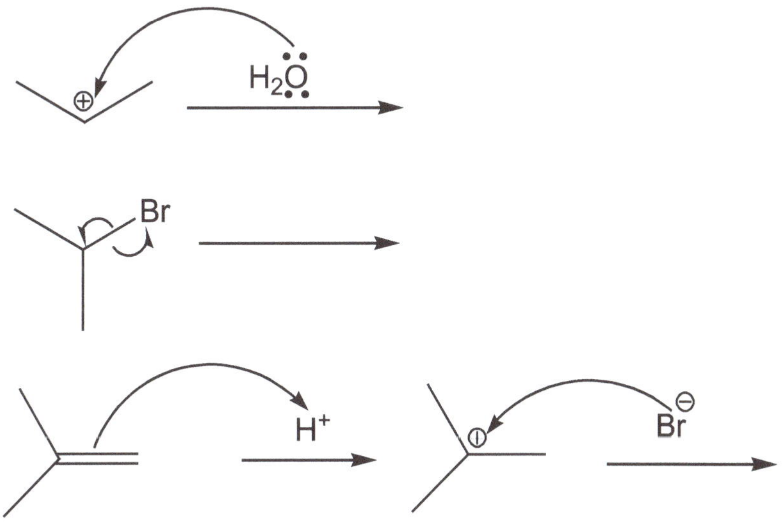

⊕ $H_2\ddot{O}$ ——→

Br ——→

H^+ ——→ ⊕ Br^- ——→

10. Write the product of the following reaction in accordance with curved arrows shown.

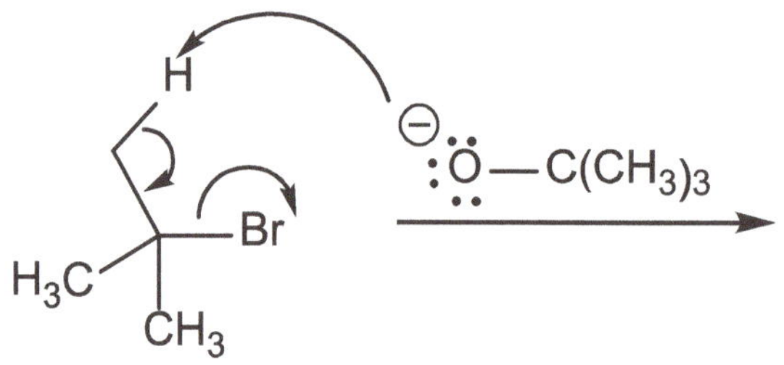

11. The pK_a values of ethanol and 2,2,2-trifluoroethanol are 15.9 and 12.4, respectively. Which one of the two is a stronger acid?

12. Identify the following as Lewis acid or Lewis base.

    a. BF_3, b. $FeCl_3$, c. $(C_2H_5)_3N$, d. $H:^-$, e. CH_3^+, f. Mg^{2+}, f. $CH_3CH{=}O$, g. CH_3OH, h. CH_3OCH_3

HYDROCARBONS-I: ALKANES AND CYCLOALKANES

Hydrocarbons, the compounds containing only carbon and hydrogen, are further divided into open chain (acyclic) and cyclic hydrocarbons based on molecular framework. Acyclic hydrocarbons are also called aliphatic (from Greek word *aleipher* meaning fat) hydrocarbons because many compounds having long chains were isolated from animal fat. These are either saturated (alkanes) or unsaturated (alkenes and alkynes). The cyclic analogs of aliphatic hydrocarbons are called alicyclic hydrocarbons. Another class of cyclic hydrocarbons, aromatic hydrocarbons differ from the rest significantly in properties and will be studied separately.

Alkanes are saturated hydrocarbons meaning they contain single bond only. They are also called paraffin due to little affinity towards other reagents (Latin, Para = little; ffin = affinity). The general molecular formula of this class of compounds is C_nH_{2n+2}. The simplest compound with n = 1 has molecular formula CH_4 and is called methane.

3.1 Structural Formula and Nomenclature

The names of all alkanes end with "*ane*". Before "*ane*", meth, eth, prop, but, pent, hex and so on are used for one, two, three, four, five and six carbon atoms, respectively (**Table 3.1**).

Table 3.1: *Homologous series of simple alkanes*

n	Compounds	Mol. Formula	Boiling points °C
1	methane	CH_4	-162 (gas)
2	ethane	C_2H_6	-89 (gas)
3	propane	C_3H_8	-42 (gas)
4	*n*-butane	C_4H_{10}	0 (gas)
5	*n*-pentane	C_5H_{12}	36 (liquid)
6	*n*-hexane	C_6H_{14}	69 (liquid)
7	*n*-heptane	C_7H_{16}	98 (liquid)
8	*n*-octane	C_8H_{18}	126 (liquid)
9	*n*-nonane	C_9H_{20}	151 (liquid)
10	*n*-decane	$C_{10}H_{22}$	174 (liquid)

A careful look at the above table showing the molecular formula of alkanes reveals that a member in the series differs in molecular formula from the member above or below it by CH_2. Thus, the compounds in Table 3.1 are said to constitute a *homologous series* of alkanes. You will observe a trend in the physical properties of the compounds belonging to one homologous series.

a. Straight-Chain Hydrocarbons

The examples of the open-chain hydrocarbons shown in Table 3.1 have *straight-chain* because the carbon atoms are connected continuously in one line; there is no branching. For example, see the structure of hexane drawn below (**Figure 3.1**).

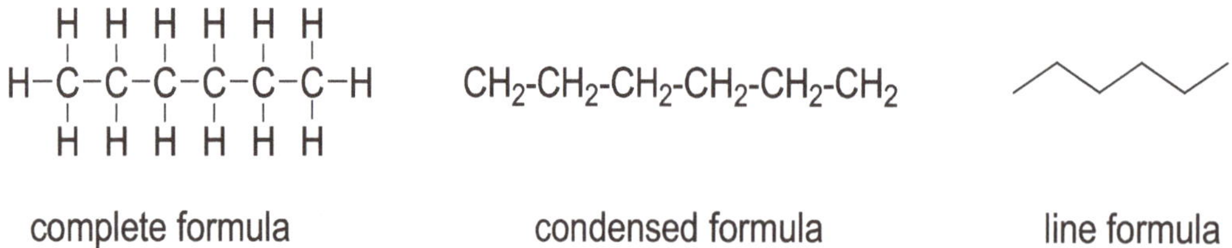

Fig 3.1: *Hexane - straight-chain skeleton.*

Problem: Look at the following line formulas and tell if they are straight-chain or branched-chain hydrocarbons.

Solution: All the line formulas are for straight-chain hydrocarbons. There is a continuous chain of carbon atoms in all of them even though they are not looking straight. In fact, they all correspond to the line formula shown in Figure 3.1.

b. Branched-Chain Hydrocarbons

When there are branches or substituents on one or more carbon atoms of the chain, the hydrocarbon is said to be *branched chain* as in the structures shown below (**Figure 3.2**).

Fig 3.2: *Open-chain hydrocarbons with branched-chain skeleton*

c. Nomenclature

To be able to name so many compounds it became necessary to devise a systematic method of nomenclature. The International Union of Pure and Applied Chemistry has developed guidelines that are known as IUPAC rules of nomenclature, which are accepted worldwide. These rules as applied to simple hydrocarbons are described below.

a. Identify the longest carbon chain in the molecule. Treat it as the parent chain.

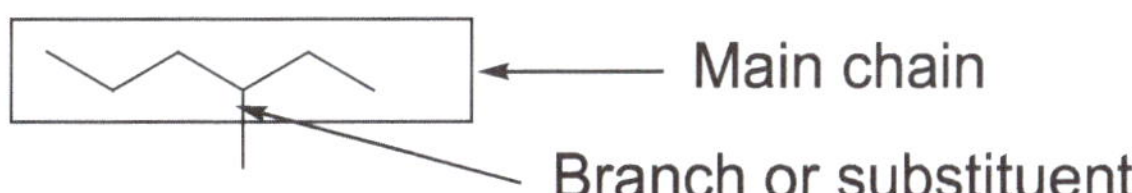

b. Identify the branches also called substituents. They will be named after the parent hydrocarbon by replacing suffix *ane* with *yl*. For example, CH_3 and C_2H_5 will be called methyl and ethyl, respectively. If the substituent is halogen (then the compound will not be hydrocarbon) it will be named like chloro, bromo, fluoro, etc.

c. To locate the position of the substituent on the parent chain, number the carbon atoms of parent chain from the side that gives minimum number to the substituent.

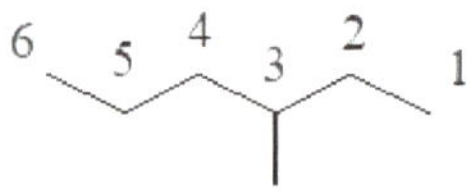

6 5 4 3 2 1

The positions of the substituents are shown in the name of the compound by Arabic numeral before the name of the parent chain. Thus, the structure above is named 3-methylhexane. Some other common branches you will encounter during your studies are shown below.

$-CH_2CH_3$	ethyl	$-HC(CH_3)CH_3$	isopropyl
$-CH_2CH_2CH_3$	propyl		
$-CH_2CH_2CH_2CH_3$	butyl	$-C(CH_3)(CH_3)H_3C$	*tert*-butyl

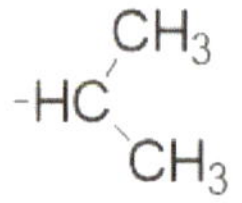

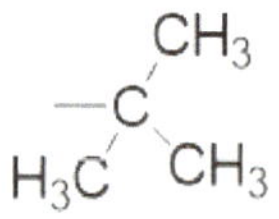

d. If there is more than one substituent, number the chain from the side giving minimum number to the first branch and arrange them in alphabetical order.

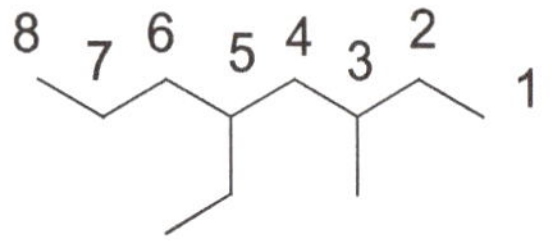

8 7 6 5 4 3 2 1

5-ethyl-3-methyloctane
(correct)

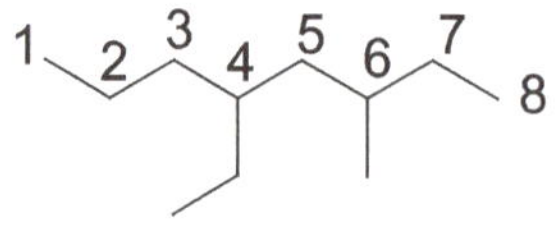

1 2 3 4 5 6 7 8

4-ethyl-6-methyloctane
(incorrect)

e. If a particular substituent appears twice or three times, prefixes such as *di-* for two, *tri-* for three, *tetra-* for four and so on are used. However, these prefixes are not taken into consideration in arranging the branches alphabetically.

3,3-dimethylhexane

3.2 Classification of Carbon and Hydrogen Atoms

In naming alkyl groups, we use prefixes *pr*- for primary, *sec*- for secondary, and *tert*- for tertiary, etc. Carbon can be classified as primary, secondary, tertiary or quaternary depending on the number of carbon atoms it is bonded to. The carbon is classified as primary or 1° if it is attached to one other carbon (**Figure 3.3**). The carbon attached to two, three and four carbon atoms are classified as secondary (2°), tertiary (3°), and quaternary (4°) carbon, respectively. The hydrogen atom attached to 1°, 2° and 3° carbon is referred to as 1°, 2° and 3° hydrogen, respectively. There is, thus, no quaternary or 4° hydrogen atom.

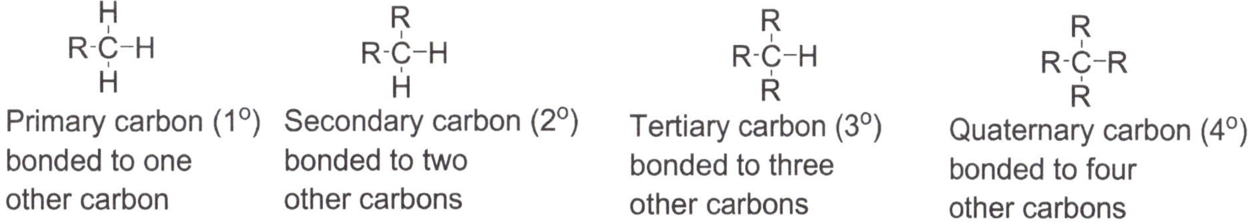

Fig. 3.3: *Classification of carbon atom.*

3.3 Isomerism in Alkanes

If you try to draw the structural formula of butane C_4H_{10}, you may do it in two different ways-by drawing a straight chain or by drawing a branched chain. Thus, as discussed in 1st chapter butane can exist as structural or constitutional/skeletal isomers (**Figure 3.4)**. You will observe both structural and geometrical isomers in cycloalkanes.

Fig 3.4: *Examples of skeletal isomerism.*

3.4 Conformations of Alkanes

It is important to know about the shape of an organic molecule as it often affects the property of the molecule. A molecule of ethane can have an infinite number of shapes due to free rotation of the one methyl group along carbon of the other methyl group. We call it carbon-carbon single bond rotation. The different possible shapes are called conformers and the property conformation. Conformers are thus stereoisomers because the bonding arrangements of the atoms are identical, they differ only in their orientation in space.

Conformers are also called rotamers because they are formed by rotation, rotation of the C-C single bond. Such rotation about single bond occurs freely because the amount of overlap of sp^3-hybridized orbitals on the two carbon atoms does not get affected by rotation of the sigma bond. Conformers are thus different from other stereoisomers in the sense that they cannot be separated because they rapidly interconvert. It has been established from studies that all rotamers are not equally stable. Some are more stable than others. The stability of the conformers depends on the orientation of the groups.

The figure below shows two conformers of ethane –staggered and eclipsed. The staggered conformation in which hydrogen atoms are as far apart as possible is the most stable conformation of ethane. It will have the least potential energy. To show the three-dimensional formula, the two-dimensional saw-horse formula is converted into the Newman projection formula (**Figure 3.5**). The "sawhorse formula is straight forward but you need to be explained about the Newman projection formula. Let's consider the staggered conformation. Assume that you are viewing from the bottom side (say front side) of the sawhorse formula. The meeting point of the three bonds in circle represents the bottom (say front) carbon. Three hydrogen atoms attached to it make the front methyl group (CH_3). The periphery of the circle represents the carbon atom of the rear methyl group.

Since conformation is the property of single bond rotation, it is not restricted only to alkanes. Compounds belonging to other groups can also exist as rotamers. See below the example of 1,2-dibromoethane ($BrH_2C\text{-}CH_2Br$) (**Figure 3.6**).

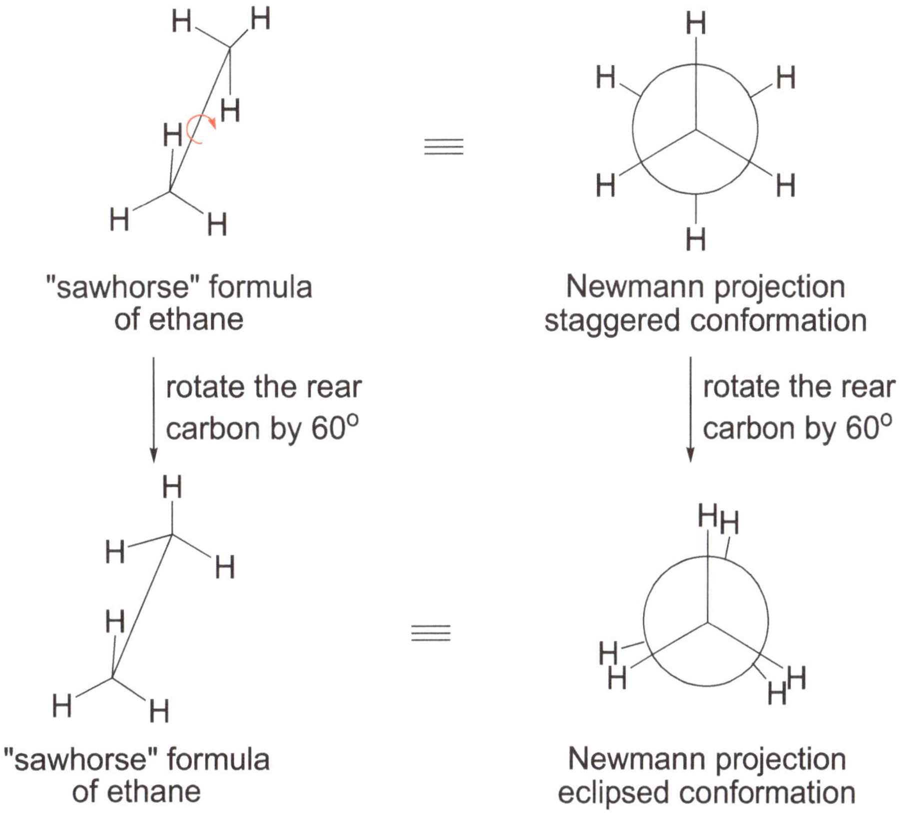

Fig 3.5: *Sawhorse and Newman projection formulas of staggered and eclipsed conformers of ethane.*

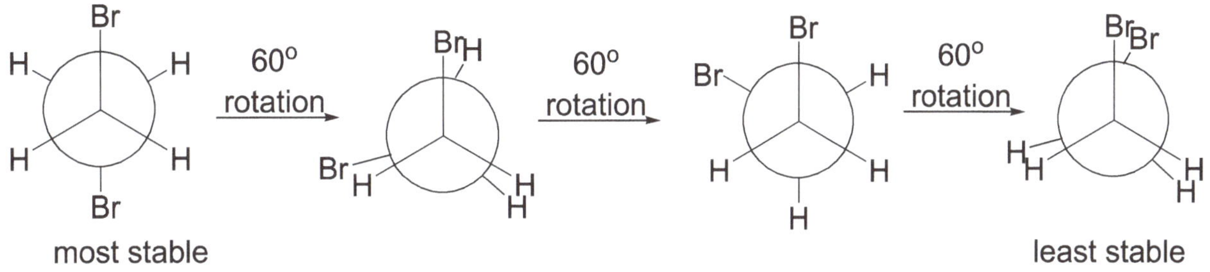

Fig 3.6: *Different conformers of 1,2-dibromoethane.*

Problem: Draw the Newman projection for different conformers of butane ($CH_3CH_2CH_2CH_3$) obtained by rotation of C2-C3 bond.

Strategy: First, draw the 1st carbon that will be C2 with one methyl group and two hydrogen atoms. Then draw the rear carbon that will be C3. Attach this carbon to a methyl group and two hydrogen atoms in such a way that the two methyl groups are *anti* to each other. Then go on rotating by 60° each time (four conformers are shown here; you can draw more).

Solution:

H H 4
1 CH3
H3C 2 3
H H
n-butane

CH3 HCH3 CH3 H3CCH3
H H 60° 60° H3C H 60°
rotation rotation rotation
H H H3CH H H HH H
CH3 H H H
most stable least stable

3.5 Occurrence

Petroleum and natural gases are the two main sources of hydrocarbons. Petroleum is also called rock oil (Latin word, petra = rock; oleum = oil). It is a complex mixture of hydrocarbons, principally alkanes and cycloalkanes. It occurs deep down on the earth as black viscous crude oil. Crude oil usually occurs together with volatile hydrocarbons called natural gases. These oils and gases are formed over a period because of decay of buried living beings.

Petroleum and natural gases are brought to the surface by drilling and pumping and refined. The various fractions of hydrocarbons (with different boiling point range) are separated by fractional distillation.

a. Fractional Distillation

It is a distillation process for separating the components of crude oils. You know that distillation of a liquid depends on the boiling point. The components with low boiling points distill first. Figure below (**Figure 3.7**) shows a fractional distillation apparatus. Crude oil is heated to vapor state and introduced in the column where a gradation in temperature. The various fractions condense and are collected. The gaseous components escape on the top where they are collected.

The boiling range and uses of the main fractions are given below.

C1—C4: -164-20 °C, natural gases: C3 and C4 are major components of cooking gas.

C5—C11: 30 to 200 °C, gasoline: C8 principal component of petrol

C11—C18: 200-370 °C: paraffin, kerosene, and diesel

C17—C22: > 400 °C: lubricating oils

C23—C34: solids mostly wax such as Vaseline

C35: solids such as asphalt

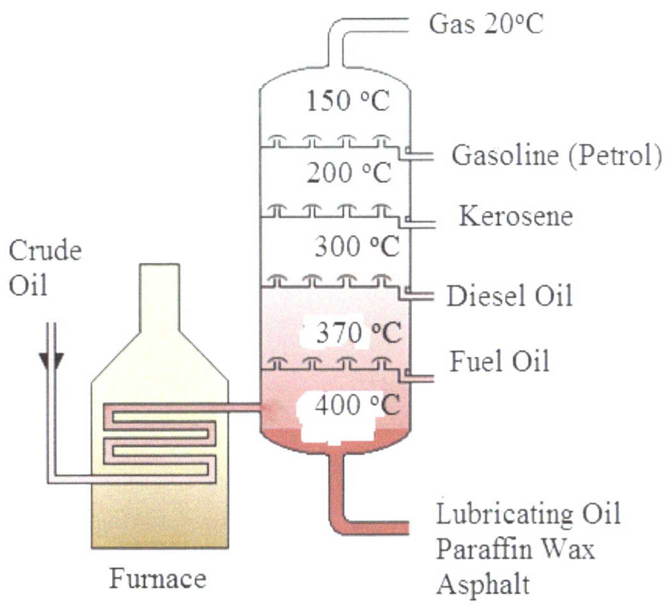

Fig. 3.7: *Apparatus for fractional distillation.*

Gasoline is the most valuable fraction as fuel and for petrochemical industries that make fibers, plastics, and other useful material. Hydrocarbons, especially those with highly branched carbon chains, burn smoothly in an automobile engine and drive the piston forward smoothly. Straight-chain hydrocarbons tend to burn violently in the engine and produce knocking because of uneven movement of the piston. The best burning petrol in an engine is 2,2,4-trimethylpentane (isooctane). Based on this, a system of rating petrol is in use that is called Octane Rating. 2,2,4-Trimethylpentane (isooctane) is arbitrarily given a rating of 100. Heptane is rated 0 because its combustion in an engine is not smooth. Thus, a fuel rated 90 has similar quality as fuel as a mixture containing 90% isooctane and 10% heptane. The octane rating of a fuel may be improved by adding tetraethyl lead, methanol, or ethanol. Lead is almost phased out because of its toxicity. Fuel stations in Botswana supply unleaded fuels with octane number 93 and 95.

b. Cracking

Cracking is the process by which high molecular weight hydrocarbons (those containing larger number of C atoms) are converted to lower hydrocarbons, which are more useful for syntheses of other compounds. Cracking is carried out either thermally (heating) or catalytically. The main catalysts for this purpose are silica and alumina. Cracking results into formation of large amounts of the hydrocarbons with lower molecular weights.

$$C_{10}H_{22} \longrightarrow \begin{cases} C_6H_{14} + C_4H_8 \\ C_5H_{12} + C_5H_{10} \\ C_4H_{10} + C_6H_{12} \end{cases}$$

3.6 Physical Properties

The physical properties of a molecule are usually referred to as its bulk properties without carrying out a chemical reaction. Such properties include physical state of the molecule, density, melting point, boiling point, refractive index, and solubility etc. These properties depend on the structure of the molecule.

a. Solubility

Hydrocarbons are insoluble in water but soluble in solvents like ether and benzene. The general rule for solubility states that "like dissolves like". Water molecules are polar whereas hydrocarbons are non-polar.

A large difference in electronegativity of oxygen and hydrogen makes the O-H bond in water polar. In hydrocarbons, C-C and C-H bonds are nearly non-polar covalent bonds. Ether and benzene molecules are also non-polar, so hydrocarbons dissolve in them.

b. Density

Hydrocarbons are less dense than water and float on its surface. This is the reason why oil spilled in the ocean remains on the surface of water.

c. Boiling Point

Boiling point increases in a homologous series with increasing number of carbon atoms for most classes of organic compounds. However, for a particular molecular weight, it decreases with increased branching of the chain. Branching makes a molecule more compact and decreases the surface area. As a result, there will be less opportunity for intermolecular attractions and consequently lesser boiling point.

36 28 10

Boiling points of pentane isomers (°C)

Alkanes have higher boiling points than their unsaturated counterparts. For example, pentane has a higher boiling point than pentene. This is due to increased repulsion in alkenes because of the presence of the double bond.

3.7 Chemical Properties

Chemical properties of organic compounds are described as their reactions with reagents. Alkanes have very limited reactivity because they are saturated molecules. The two main reactions of alkanes are with a. oxygen (combustion), and b. halogens (halogenation) which fall under substitution reactions.

a. Combustion

Combustion is the most important reaction of alkanes as it provides both light and heat energy. Heat energy is convertible to several other forms such as the movement of limbs, machines, and several others. In the presence of excess oxygen, alkanes burn with a clear flame. How to account the heat energy produced is taught in thermochemistry.

$$CH_4 + 2\,O_2 \rightarrow CO_2 + 2H_2O + \text{heat}$$

$$2C_4H_{10} + 13\,O_2 \rightarrow 8\,CO_2 + 10\,H_2O + \text{heat}$$

You can see from the above reaction equation that the hydrogen atoms in alkanes are substituted by oxygen atoms. Thus, combustion is an oxidation reaction. If oxygen present is insufficient, carbon monoxide can also be produced because of combustion.

$$2CH_4 + 3\,O_2 \rightarrow 2\,CO + 4H_2O$$

b. Halogenation

Halogenation reactions are examples of substitution reactions. One or more hydrogen atoms of hydrocarbons are replaced by halogen (Cl or Br) atoms. The reaction is usually carried out by exposing a mixture of an alkane and halogen to light. *Remember that the organic product(s) of halogenation reaction would no longer be a hydrocarbon.* It would contain halogen atom(s) besides carbon and hydrogen atoms. These products are medicinally useful and called haloalkanes. A haloalkane, 1-bromo-1-chloroethane is used as general anesthetic. Halogen atom(s) will be treated as branch on main chain in naming such compounds. Inorganic products are not treated as main products of organic reaction because the primary purpose of studying chemical properties of organic compounds is to transform them into another organic compound.

$$CH_4 + Cl_2 \xrightarrow{\text{light}} CH_3Cl + HCl$$

The reaction is called chlorination if halogen is chlorine and bromination if halogen is bromine. Monohalogenation (replacement of only one hydrogen atom) in such reactions is rarely possible. It is difficult to stop the reaction at monohalogenation step and a mixture of products is obtained. Therefore, the halogenation of alkanes is not a very useful reaction for the preparation of monohaloalkanes. If an excess of halogen is present, the reaction goes on until all hydrogen atoms are substituted. For example, ethane reacts with excess chlorine in the presence of light to form hexachloroethane.

$$C_2H_6 + \text{excess } Cl_2 \rightarrow C_2Cl_6 + 6\ HCl$$

In the case of monohalogenation as well, more than one product can be obtained in the case of hydrocarbons that have hydrogen atoms that are non-equivalent. Look at the structure of 2,3-dimethylbutane. It has twelve hydrogen atoms at four methyl groups that are equivalent. Another two hydrogen atoms at C-2 and C-3, which are equivalent, are not equivalent to methyl group hydrogen atoms. So, in this case, monobromination will give two isomeric bromobutanes.

$$(H_3C)_2CH{-}CH(CH_3)_2 + Br_2 \xrightarrow{\text{light}} (H_3C)_2CH{-}CH(CH_3){-}H_2C{-}Br + (H_3C)_2CH{-}C(Br)(CH_3)_2$$

Problem: Write the structure of all possible monochlorination products that may form on chlorination of propane.

$$H_3C{-}CH_2{-}CH_3$$

Solution: The two terminal methyl groups in this compound are equivalent. Whichever hydrogen you substitute, the same product 1-chloropropane will be formed. If you substitute the hydrogen atom on middle carbon, 2-chloropropane will be formed. Thus, two monochlorination products are possible.

$$H_3C{-}CH_2{-}CH_3 + Cl_2 \xrightarrow{\text{light}} H_3C{-}CH_2{-}CH_2Cl + H_3C{-}CHCl{-}CH_3$$

1-chloropropane 2-chloropropane

c. Mechanism of Halogenation of Alkanes

Halogenation of alkanes is a three-step mechanism - initiation, chain propagation and chain termination involving radicals. Therefore, the halogenation reactions of alkanes are said to occur by *radical mechanism*. The first two steps involve formation of intermediates with an unpaired electron in the valence shell. These intermediates are called free-radicals or simply radicals. The radicals react either with neutral molecules or with another radical.

Initiation: It involves homolytic cleavage of the Cl-Cl bond by light energy. Homolytic cleavage, as described in Section 2.5, is the cleavage of a covalent bond in which an electron pair is shared equally between the two atoms joined by that covalent bond. Thus, each of the chlorine atoms gets away with one electron forming two chlorine radicals that are shown in equation below with the symbol of chlorine atom and a "dot" over it which represents an unshared electron.

$$Cl{-}Cl \xrightarrow{\text{light}} {}^{\bullet}Cl + {}^{\bullet}Cl$$

Propagation: It involves the reaction of a radical with a neutral molecule forming a new radical and a neutral product (the main reaction product here).

$$CH_4 + {}^{\bullet}Cl \longrightarrow {}^{\bullet}CH_3 + HCl$$

$${}^{\bullet}CH_3 + Cl{-}Cl \longrightarrow Cl{-}CH_3 + {}^{\bullet}Cl$$

Termination: It involves a reaction between two identical or different radicals forming molecules. Now you can see that *the main product of the reaction, chloromethane, is formed during propagation and termination steps*. The formation of a small amount of ethane in the halogenation of methane supports the free radical mechanism.

$$\dot{C}H_3 + \dot{C}H_3 \longrightarrow H_3C{-}CH_3$$

$$\dot{C}H_3 + \dot{C}l \longrightarrow Cl{-}CH_3$$

3.8 Cycloalkanes

a. Structural Formula and Nomenclature

The cyclic or ring analogs of the open chain alkanes shown above are called cycloalkanes. The prefix *cyclo* is included in their names to indicate the presence of the ring. Since the two ends of chain will be joined in such compounds, they will have two hydrogen atoms less than their open chain counterpart. For example, propane has the molecular formula C_3H_8, so cyclopropane will have molecular formula C_3H_6.

$$\underset{\text{propane}}{H_2C{-}CH_2{-}CH_2\ (H,\ H)} \xrightarrow{-2H} \underset{\text{cyclopropane}}{\text{cyclo-}(CH_2)_3}$$

Thus, the general molecular formula of cycloalkanes is C_nH_{2n}. The study of cycloalkanes is of immense importance in organic chemistry because such rings are often present in compounds occurring in nature. Cyclopropane obviously is the smallest cycloalkane. Some other simple examples are cyclobutane (C_4H_8), cyclopentane (C_5H_{10}) cyclohexane (C_6H_{12}), cycloheptane (C_7H_{14}), and cyclooctane (C_8H_{16}) (**Figure 3.8**).

Fig. 3.8: *Structures of some simple cycloalkanes.*

Cycloalkanes are named in similar way as open-chain alkanes considering cycle as the main unit and prefix *cyclo* is used before the name of the parent hydrocarbon. For example, a ring bearing four carbon atoms is called cyclobutane. If the cycle has a branch, always start numbering the cycle from the carbon that bears the branch. If there is more than one branch, then give alphabetical priority to the branches and number from the side that gives minimum number to other branches (**Figure 3.9**).

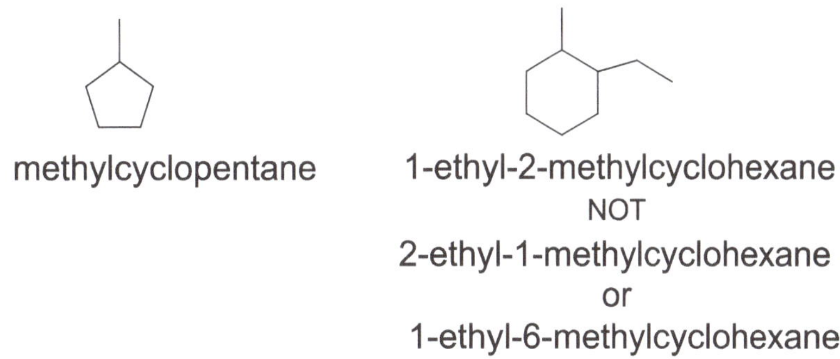

Fig.: 3.9: *Naming branched cycloalkanes.*

b. Isomerism in Cycloalkanes

Cycloalkanes show both constitutional and geometrical isomerism. Constitutional isomerism of skeletal (**Figure 3.10**) or positional types (**Figure 3.11**) is possible due to completely different skeletons with the same molecular formula or due to different positions of branches on the ring.

You must keep in mind that the compounds belonging to different classes can also be isomers to each other. To be of the same class is not a necessary condition for being isomers.

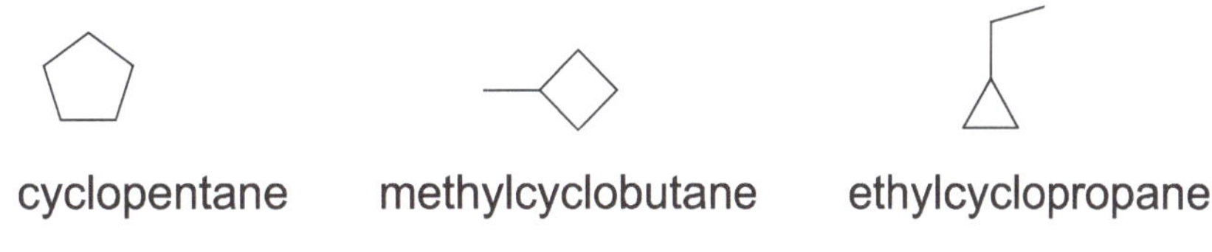

Fig. 3.10: *Structural isomers (skeletal) of cyclopentane.*

1,2-dimethylcyclohexane 1,3-dimethylcyclohexane 1,4-dimethylcyclohexane

Fig. 3.11: *Structural (positional) isomers of dimethylcyclohexane.*

As stated earlier, cycloalkanes differ from alkanes structurally in term of “chain” and “cycle or ring”. This difference gives a special structural feature to cycloalkanes. The cycloalkane molecules are “rigid” in terms of rotation of single bonds. The carbon-carbon single bonds in alkanes can rotate freely but not in cycloalkanes. If there are two substituents on the ring, they may be on the same side relative to the plane of the ring, or they may be on opposite sides. This leads to geometrical isomerism. If the two substituents are in the same plane, the isomer is called *cis*-isomer, if they are on opposite sides it is called *trans*-isomer (**Figure 3.12**).

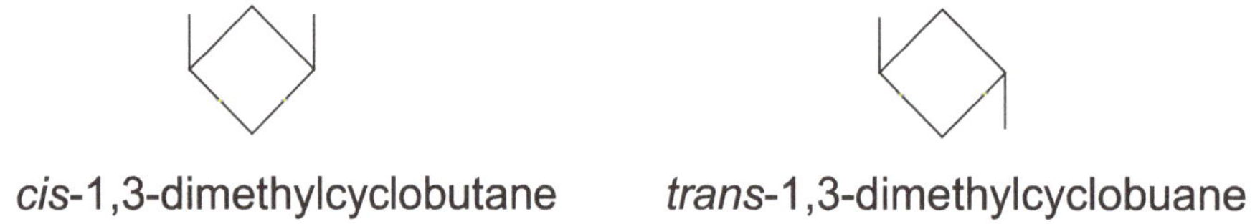

Fig. 3.12: *Geometrical isomers of 1,3-dimethylcyclohexane.*

c. Properties

The physical properties of cycloalkanes will show similar trend as observed in alkanes. For example, the boiling point of cycloalkanes increases with increasing ring-size. Thus, cyclopropane has a lower boiling point than cyclobutane, which has a lower boiling point than cyclopentane. Chemical properties are also like those of alkanes. Thus, cycloalkanes undergo combustion and halogenation reactions.

d. Conformations of Cycloalkanes

The smallest cycloalkane, cyclopropane is a planar molecule. The C-C-C bond angle is only 60° that is much lower than the normal tetrahedral angle of 109.5° (**Figure 3.13**). In compensation, the H-C-H bond angle is 120° which is much larger than the normal angle for sp^3-hybridized carbon. The hydrogen atoms lie above and below the plane of the ring, and hydrogen atoms on adjacent carbons are eclipsed.

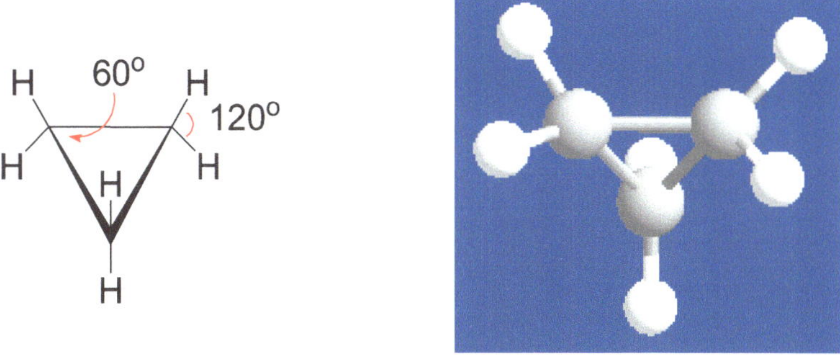

Fig. 3.13: *Conformation of cyclopropane.*

Cyclobutane and cyclopentane are nonplanar and have slightly puckered conformations (**Figure 3.14**). This puckering allows the molecule to adopt the most stable conformation with the least strain energy. Puckering makes the C-C-C bond angles a bit smaller than they would be if the molecules were planar. At the same time, the puckering reduces the eclipsing of the adjacent hydrogen atoms.

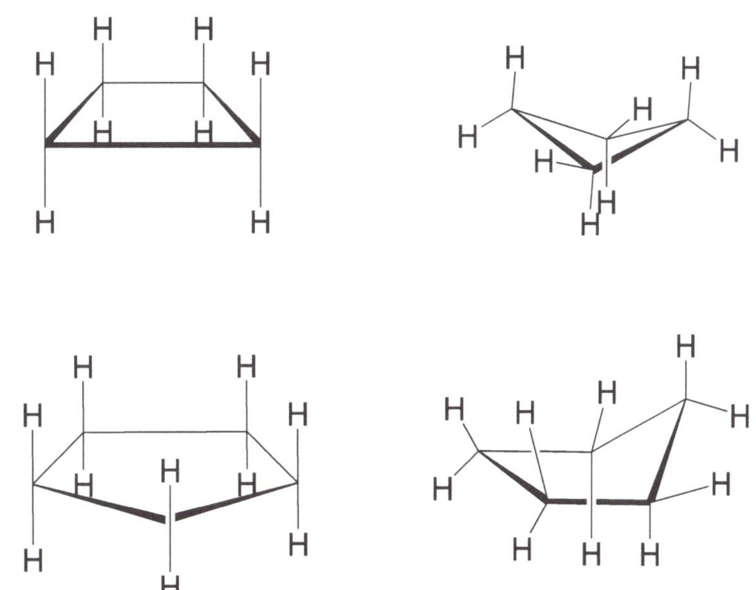

Fig. 3.14: *Conformations of cyclobutane and cyclopentane.*

Six-membered rings have been studied in detail because of their widespread occurrence in nature. Cyclohexane ring occurs in many compounds of pharmaceutical interest. You will be surprised to learn that cyclohexane is not a flat molecule. If cyclohexane were flat (planar), the internal C-C-C bond angles would be that of a regular hexagon, 120°. This bond angle is a bit much larger than the normal tetrahedral angle of 109.5°. The resulting strain prevents cyclohexane from being a planar (flat) molecule. Cyclohexane adopts a strain-free three-dimensional shape called a chair conformation (**Figure 3.15**) because of similarity of its shape with a chair. In the chair conformation, all C-C-C bond angles are 109.5° and all the hydrogen atoms on adjacent carbons are perfectly staggered. Therefore, the chair conformation of the cyclohexane is the most stable conformation.

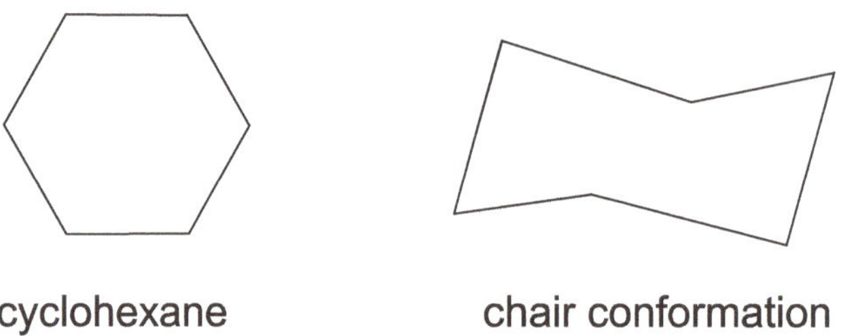

Fig. 3.15: *Chair conformation of cyclohexane.*

Due to puckering to chair form, the orientation of all hydrogen atoms does not remain same. There are two kinds of positions for hydrogen atoms on the ring – *axial* positions for six hydrogen atoms and *equatorial* positions for six hydrogen atoms (**Figure 3.16**). Among the six axial hydrogen atoms, three axial hydrogen atoms (say one each on C1, C3 and C5) lie above the plane of the ring and three axial hydrogen atoms (say one each on C2, C4 and C6) lie below the plane of the ring. The six equatorial hydrogen atoms lie on the equator of the ring almost in its plane. Different chair conformations of cyclohexane readily interconvert at room temperature. As a result of interconversion, which is called "ring-flip", all axial hydrogen atoms become equatorial and all equatorial hydrogen atoms become axial (see the figure below).

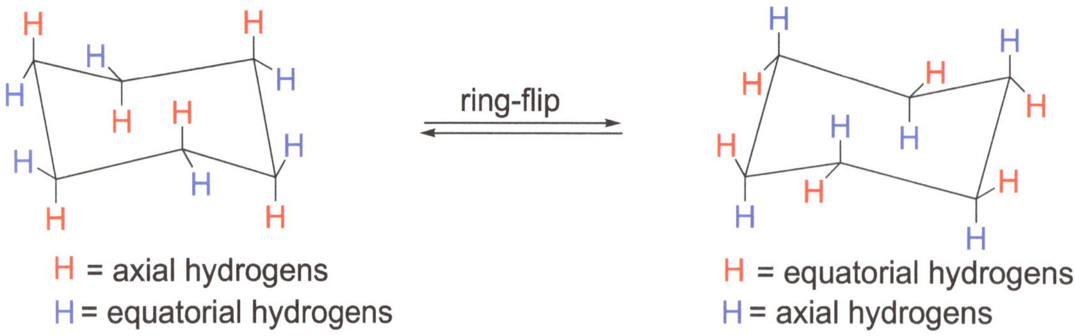

Fig. 3.16: *Ring-flipping in chair conformation of cyclohexane.*

In the beginning, you may feel difficulty in drawing the chair shape. You can draw it as shown in the scheme below (**Figure 3.17**). However, it is advisable to you to make model of the chair conformation which will give you a better understanding of three-dimensional structure.

Fig. 3.17: *Drawing chair conformation of cyclohexane.*

Now let's think about a monosubstituted cyclohexane such as methylcyclohexane. You may assume that it will have two isomers because methyl group can occupy either axial or equatorial position. But this is not true. There is only one methylcyclohexane because the ring flips easily at room temperature as in case of cyclohexane, and axial and equatorial methyl groups change their positions. However, the two chair conformations of methylcyclohexane are not equally stable. The chair conformation with methyl group in equatorial position has 7.6 kJ/mol less energy than the chair conformation with methyl group occupying axial position. It means that the former conformation is more stable by 7.6 kJ/mol energy than the latter one. The methyl cyclohexane with methyl group occupying axial position is less stable because the methyl hydrogen atoms become too close to axial hydrogen atoms in that plane of the cyclohexane ring. This proximity of the axial methyl group with axial hydrogens is called 1,3-diaxial interaction (**Figure 3.18**). It creates steric strain in the molecule making it less stable. This is true for all monosubstituted cyclohexane rings. It means that the chair conformation of monosubstituted cyclohexane rings with substituent on equatorial position will be more stable than the other with substituent on axial position.

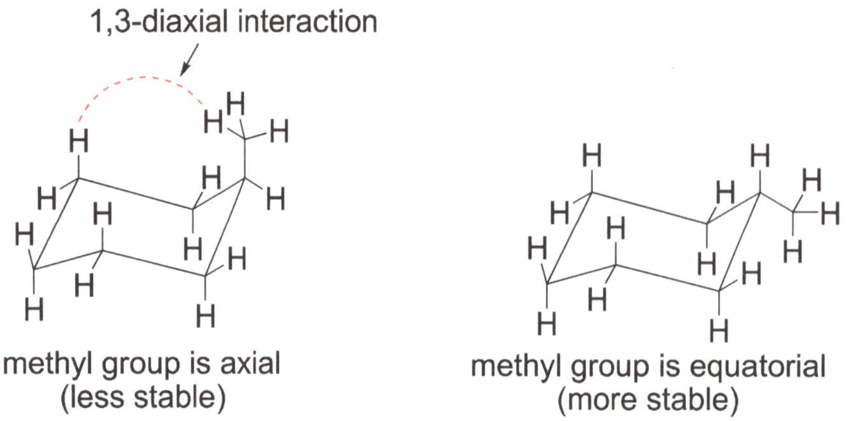

1,3-diaxial interaction

methyl group is axial
(less stable)

methyl group is equatorial
(more stable)

Fig. 3.18: *1,3-Diaxial interaction in cyclohexane with axial methyl group.*

SUMMARY

Alkanes and cycloalkanes are saturated hydrocarbons or paraffin. Their main sources are petroleum and natural gases. They are named according to the IUPAC rules. Alkanes, because they are saturated, have little affinity for chemical reactions. They undergo combustion (oxidation) to form carbon dioxide and water. This is an exothermic reaction and hence heat energy is released during the reaction. Another common chemical reaction of alkanes is substitution of one or more hydrogen atoms by one or more halogen atoms. This reaction occurs in the presence of light by a free-radical mechanism. The free-radical mechanism occurs in three steps – initiation, propagation, and termination. Alkanes and cycloalkanes show structural isomerism of skeletal or positional types. Cycloalkanes also exist as geometrical isomers because of the rigid ring structure. The latter is also responsible for giving different conformational structures to cycloalkanes. Cyclopropane is a planar molecule but cyclobutane and cyclopentane are nonplanar and have slightly puckered conformations. The cyclohexane exists mostly in chair-form. In open-chain alkanes, however, free-rotation of carbon-carbon bond is possible. This leads to several conformations that are readily interconvertible. The most stable conformation for ethane is staggered conformation.

EXERCISES

1. Write the condensed structure formula, detailed structural formula and line formula of the alkanes and cycloalkanes having following molecular formula.

   a. C_3H_8, b. C_4H_{10} (branched chain), c. C_5H_{12} (straight chain), d. C_5H_{10} (cycloalkane with four carbon atoms in ring), e. C_5H_{10} (cycloalkane with three carbon atoms in ring)

2. Name the following alkanes and cycloalkanes according to the IUPAC rules.

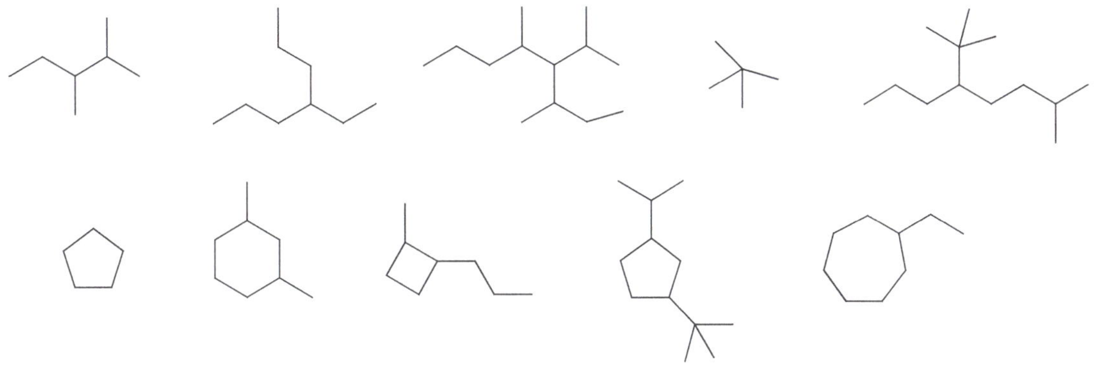

3. Write the condensed structural formula and molecular formula of compounds corresponding to line formulas in 2.

4. Write the condensed structural formula and line formula for the following compounds.

   a. 3,3-dimethyl-4-ethylhexane
   b. 3-ethyl-2-methyloctane
   c. 4-isopropyl-3-methylheptane
   d. 2-methylhexane
   e. 1-ethyl-3-methylcyclohexane
   f. bromocyclobutane
   g. 1,1-dichlorocyclopentane
   h. 3-ethyl-2,2-dimethylhexane

5. Write structural formulas for all possible structural isomers of an alkane with molecular formula C_5H_{12}.

6. Identify the carbon and hydrogen atoms in the following compounds as primary, secondary, tertiary, and quaternary.

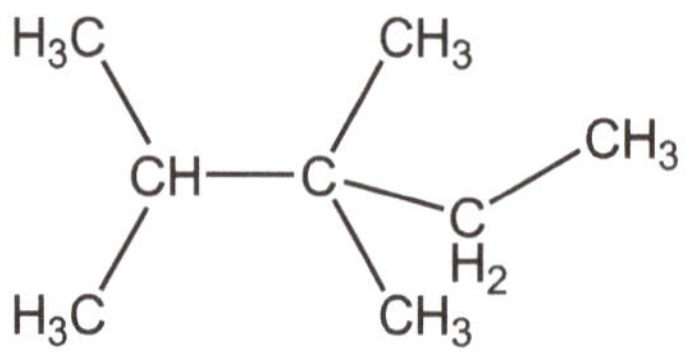

7. Write all possible cycloalkane structural isomers with molecular formula C_4H_8 and C_5H_{10}.

8. Alkanes do not exist as *cis-trans* isomers but cycloalkanes do. Briefly explain.

9. Explain the difference between *cis-trans* isomerism and conformational isomerism.

10. Draw two Newman projections of propane by looking down C_1-C_2 bond.

11. Describe briefly how petroleum was formed.

12. Define the following terms and their application in petroleum industry.

    a. Cracking, b. octane rating, c. fractional distillation

13. A certain gasoline has an octane rating of 91. What does it mean?

14. The boiling point of *n*-pentane is 36 °C while its isomer 2,2-dimethylpropane boils at 10 °C. Explain.

15. Briefly explain why alkanes are insoluble in water but alcohols like methanol and ethanol are soluble in water.

16. Arrange the following series of compounds in order of their increasing boiling point (lowest b. p. first).

a.

b.

17. Write a balanced chemical equation for combustion of 2,2,3-trimethylpentane.

18. How many monochlorination products will be obtained from the reaction of following alkanes/cycloalkanes with chlorine in the presence of light? Write the structure of product(s) in each case.

19. The reaction of ethane with bromine in the presence of light gives butane in trace amount together with bromoethane. Explain the formation of butane by writing the mechanism of bromination of ethane.

20. Draw the structural formula for the preferred conformation of following cyclohexane derivatives.

    a. cyclohexane
    b. trans-1,2-dimethylcyclohexane
    c. cis-1-chloro-3-ethylcyclohexane
    d. chlorocyclohexane

21. Using molecular model, identify the following cyclohexane derivatives as *cis* or *trans* isomer.

CH_3 CH_3 CH_3 CH_3 H_3C CH_3

22. Identify the following compounds as conformers, structural isomers, geometrical isomers or identical.

i. CH_3 H H H H CH_3 and CH_3 H_3C H H H H ii. and

iii. Br H H H H Br and Br H H H Br H iv. Cl Cl and Cl Cl

HYDROCARBONS-II: ALKENES AND ALKYNES

4.1 Structure and Nomenclature

Alkenes are unsaturated hydrocarbons also known as olefins. They contain at least one carbon-carbon double bond. The carbon atoms in C=C are sp^2-hybridized with bond angles of 120°. Also, one of the double bonds is a *sigma* bond and the other is a *pi*-bond. The general molecular formula of this class of compounds is C_nH_{2n}. This general molecular formula is identical to the general molecular formula of simple cycloalkanes. The simplest compound with n = 2 has molecular formula C_2H_4 and is called ethene or ethylene. The names of all the alkenes end with *"ene"*. Prefixes like eth, prop, but, pent and so on are used for two, three, four and five carbons (**Table 4.1**), respectively, in the parent chain as used for alkanes. Analogous to cycloalkanes, cyclic analogs of alkenes are called cycloalkenes. Their general molecular formula C_nH_{2n-2} has two hydrogen atoms less than the open-chain alkenes. Later, you will notice that the general molecular formula of cycloalkenes and open-chain alkynes are same.

Table 4.1: *Homologous series of simple alkenes*

n	Compounds	Mol.Formula	n	Compounds	Mol. Formula
2	Ethene	C_2H_4	7	Heptene	C_7H_{14}
3	Propene	C_3H_6	8	Octene	C_8H_{16}
4	Butene	C_4H_8	9	Nonene	C_9H_{18}
5	Pentene	C_5H_{10}	10	Decene	$C_{10}H_{20}$
6	Hexene	C_6H_{12}	11	Undecene	$C_{11}H_{22}$

Alkynes are also unsaturated hydrocarbons. They contain at least on carbon-carbon triple bond. Carbon-carbon triple bond is *sp*-hybridized with linear geometry. Remember that two carbon-carbon bonds in the triple bond are pi bonds and the third one is sigma bond. They greatly resemble alkenes in their chemical properties. The general molecular formula of this class of compounds is C_nH_{2n-2}. The simplest compound with n = 2 has molecular formula C_2H_2 and is called ethyne. Its common name is acetylene that is used for wielding.

The names of all alkynes end with *"yne"*. The smallest cycloalkyne known is cyclooctyne. Due to linear geometry of carbon atoms involved in triple bond formation, formation of smaller cycloalkyne rings is not possible. The rules for naming alkynes are same as those for alkenes (**Table 4.2**).

Table 4.2: *Homologous series of simple alkynes.*

n	Compounds	Mol. Formula	n	Compounds	Mol. Formula
2	Ethyne	C_2H_2	7	Heptyne	C_7H_{12}
3	Propyne	C_3H_4	8	Octyne	C_8H_{14}
4	Butyne	C_4H_6	9	Nonyne	C_9H_{16}
5	Pentyne	C_5H_8	10	Decyne	$C_{10}H_{18}$
6	Hexyne	C_6H_{10}	11	Undecyne	$C_{11}H_{20}$

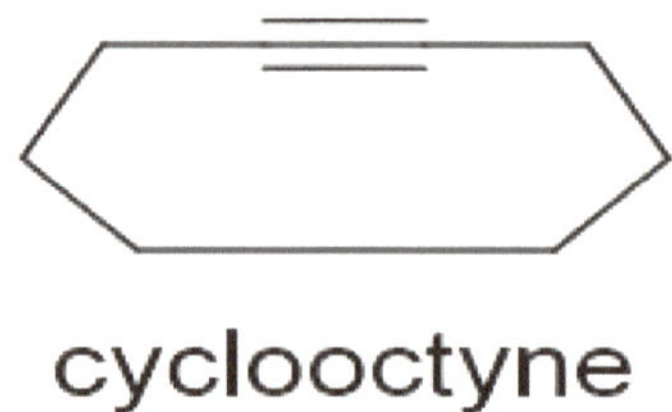

cyclooctyne

1. Similar to alkanes, first you must identify the parent or main chain. In alkenes or alkynes containing four or more carbon atoms, the position of carbon-carbon double must be indicated. The carbon chain is numbered from the side that gives minimum number to the first carbon attached to the carbon-carbon double bond.

$H_2C{=}CHCH_2CH_3$	$H_3C{-}\underset{H}{C}{=}\underset{H}{C}{-}CH_3$	$HC{\equiv}CCH_2CH_3$	$H_3CC{\equiv}CCH_3$
1-butene	2-butene	1-butyne	2-butyne

2. In branched-chain unsaturated hydrocarbons, the chain containing the double or triple bond is considered as the main chain (even if it is not the longest one). In the example shown below, the longest possible chain has nine carbon atoms. But the chain with eight carbon atoms is taken as main chain because it contains the carbon-carbon double bond. The chain is numbered from the side that gives minimum number to the double bond. If a chain has both double bond and an alkyl substituent (branch), give the minimum possible number to the double bond.

3-propyl-2-octene
NOT
6-propyl-6-octene

3. In naming a branched cycloalkene, the numbering of carbon always starts from the carbon attached with the double bond and not from the carbon bearing the branch as in cycloalkanes. The second carbon will always be the other carbon atom attached to the double bond. The numbering is started from that C of the two attached by the double bond which gives the lower number to the substituent as shown in the examples below.

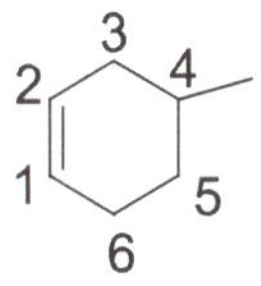

4-methyl cyclohexene
NOT
5-methylcyclohexene

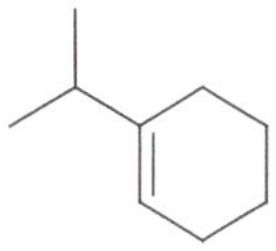

1-isopropylcyclohexene
NOT
2-isopropylcyclohexene

4. The compounds containing both carbon-carbon double and triple bonds are known as "enyne" (from ene + yne) and not "ynene". The parent chain is named from the side that gives minimum number to the multiple bond whether it is double or triple bond. If both double bond and triple bond are equidistant, double bond gets higher priority (minimum number).

$H_3C{-}C{\equiv}C{-}CH_2CH_2CH{=}CHCH_3$
oct-2-en-6-yne

$HC{\equiv}C{-}CH_2{=}CHCH_3$
pent-3-ene-1-yne

5. **Allylic and benzylic carbon and hydrogen atom**: Sometimes common names are used to denote some types of carbon and hydrogen atoms present in olefinic systems. For example, the carbon and hydrogen atoms of a CH_2 attached to C=C are called allylic carbon and allylic hydrogen. However, if the C=C to which the CH_2 is attached is present in benzene ring, then the carbon and hydrogen are referred to as benzylic carbon and benzylic hydrogens. A $H_2C{=}CH$ system is referred to as vinyl group in alkene chemistry. The presence of allylic, benzylic or vinylic system in compounds imparts special reactivity to such compounds.

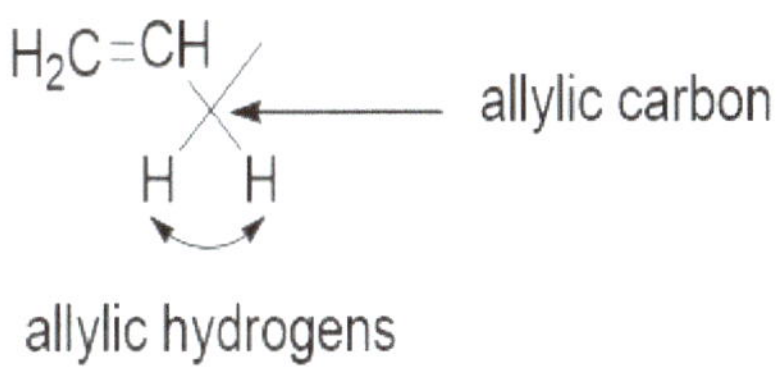

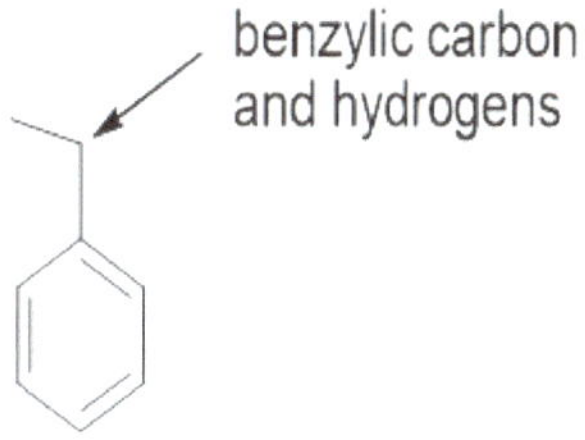

$H_2C{=}CH$
H H
allylic carbon
allylic hydrogens
benzylic carbon and hydrogens

4.2 Isomerism

As you saw in the nomenclature, a double bond in alkenes having four or more carbon atoms can have different positions on the chain. As a result, such alkenes can exist as structural isomers (positional isomers).

$CH_2{=}CH\text{-}CH_2\text{-}CH_3$ — 1-butene

$CH_3\text{-}CH{=}CH\text{-}CH_3$ — 2-butene

structural isomers

Alkenes can also be isomeric to cycloalkanes because they have same general molecular formula. For example, 1- and 2-butenes are structural isomers of cyclobutane.

$CH_2{=}CH\text{-}CH_2\text{-}CH_3$ — 1-butene

$CH_3\text{-}CH{=}CH\text{-}CH_3$ — 2-butene

$H_2C{-}CH_2$, $H_2C{-}CH_2$ (ring) — cyclobutane

structural isomers

Alkenes also exhibit geometrical isomerism. It arises due to rigidity of the carbon-carbon double bond. *Alkenes having different atoms or groups attached to the doubly bonded carbon atoms will have two isomers*. In one of them, the identical groups are on same side of the double bond; these are called *cis* isomers. In another isomer, identical groups will be on the opposite side of the double bond and are called *trans* isomers. If either of the carbon atoms attached to the double bond has identical atoms or groups, the compound cannot exhibit geometrical isomerism (**Figure 4.1**).

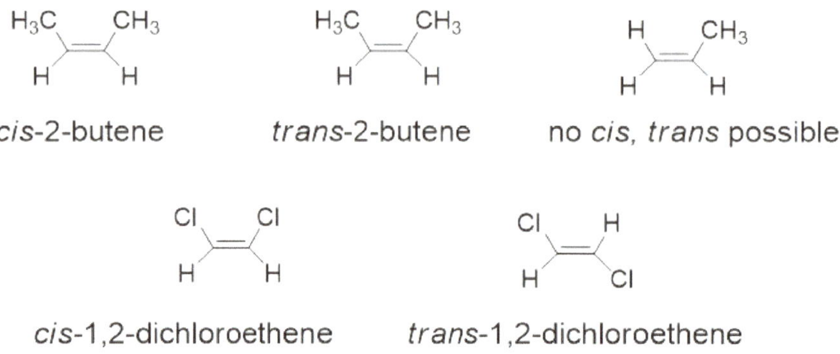

Fig. 4.1: *Cis and trans isomerism.*

E-Z System of Nomenclature

In the examples of geometrical isomers shown above, the groups attached to both carbons of C=C are same. For example, in 1,2-dichloroethene, both carbon atoms of C=C have one hydrogen atom and one chlorine atom. This compound and others like it, can easily be named as *cis* or *trans* depending on whether the two chlorines are on same side of the double bond or on the opposite side. Look at the example below in which the groups attached to C=C are not only of two types but are all different. We do not or rather cannot name it as *cis* or *trans*.

H CH_3 / Br CH_2CH_3 Br CH_3 / H CH_2CH_3

For naming such compounds, a convenient *E-Z* system of nomenclature has been devised. In this system, the groups/atoms attached to each carbon are prioritized as 1 (higher) or 2 (lower) according to certain rules (see *R-S* system of nomenclature for priority rules in chapter 5). If the groups with identical priority are on same side of the C=C, the isomer is called *Z*-isomer (from the German zusammen meaning together). If the groups with identical priority are on opposite side of the C=C, the isomer is called *E*-isomer (from the German entgegen meaning opposite). In the example shown above, one carbon of C=C has a bromo group and a hydrogen atom. Among these two, bromo has higher priority (1) and H has lower priority (2) based on their atomic numbers. Other carbon has a methyl group and an ethyl group. Among these two, ethyl group has higher priority (1) and methyl group has lower priority (2) based on what atoms are attached to carbon atoms joined to C=C. In CH_3 group, the C is attached to three H atoms while in CH_2CH_3 group the carbon is attached to two H and one C atom. Hence the ethyl group has priority over methyl group. See the classification of these isomers as *E* and *Z* below (**Figure 4.2**).

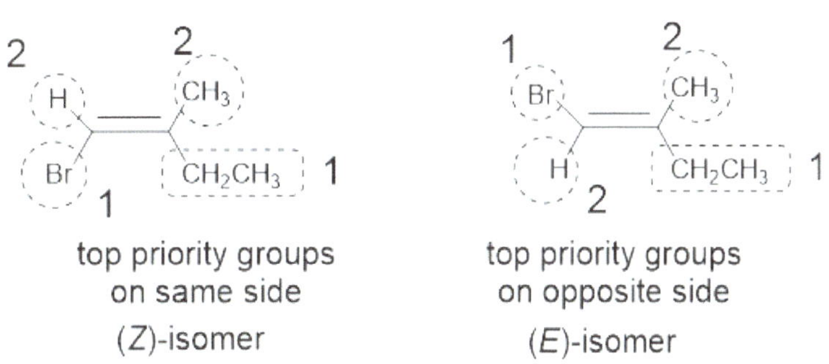

Fig. 4.2: *Identifying E and Z isomers.*

4.3 Physical Properties

The physical properties of alkenes and alkynes resemble alkanes. The lower members in the homologous series are gases while higher members are liquids. The boiling point thus increases gradually with increasing chain length.

Alkenes have lower boiling points than its saturated counterparts. For example, pentene has lower boiling point than pentane. This is due to increased repulsion in alkenes because of the presence of the double bond.

They are insoluble in water or but soluble in non-polar organic solvents such as ether or benzene.

4.4 Chemical Properties of Alkenes

Alkenes undergo a lot more reactions than alkanes. This is because of the presence of π-electrons in the double bond, which are loosely bound.

a. Combustion

Alkenes, like alkanes, are easily oxidized to carbon dioxide and water when burnt in excess oxygen. The flames are sooty unlike alkanes because of a higher C to H ratio.

$$C_nH_{2n} + 3/2\ O_2 \longrightarrow n\ CO_2 + n\ H_2O + \text{heat}$$

b. Addition Reactions of Alkenes

The most common reactions of alkenes are addition reactions. In such reactions, the atoms or groups of the reagent are added to carbon atoms attached by double bond yielding products in which these carbon atoms have single bond. The reagents might be symmetrical such as hydrogen (H_2), chlorine (Cl_2) or bromine (Br_2) or unsymmetrical such as hydrogen halides, (H-X) and water (H-OH), etc., and so the alkenes or alkynes.

Addition of hydrogen (Reduction): Preparation of alkanes

Addition of hydrogen to alkenes, also known as hydrogenation, is an example of *reduction* reaction of alkenes. The reaction occurs in the presence of a catalyst. Some commonly used catalysts are metals like Pt, Pd, and Ni. This reaction is of great value and is used in production of margarine from oil.

$$\text{1-butene} + H_2 \xrightarrow{\text{Pt or Ni}} \text{butane}$$

Addition of halogens: Preparation of dihaloalkanes

These reactions take place at room temperature without any catalyst or necessity of light. Note again that the product of halogen addition is no more a hydrocarbon. Furthermore, two halogen atoms are added to alkene molecules; hence the products are dihaloalkanes. These reactions do not need any control because the addition will stop if there is no more unsaturation (double bond) in the alkene molecule. Thus, only one halogen molecule can be added to one alkene molecule. The reaction of 1-butene with chlorine, thus, gives 1,2-dichlorobutane. Similarly, *trans*-2-pentene reacts with bromine to form 2,3-dibromopentane.

Bromine-water solution is used to test the presence of unsaturation in alkenes and cycloalkenes. Such hydrocarbons decolorize the orange color bromine solution.

$$\text{1-butene} + Cl_2 \longrightarrow \text{1,2-dichlorobutane}$$

$$\textit{trans}\text{-2-pentene} + Br_2\text{ (aq)} \longrightarrow \text{2,3-dibromopentane}$$

Addition of hydrogen halides: Preparation of monohaloalkanes

The common examples are addition of HCl and HBr to alkenes. The reactions of alkenes with hydrogen halides lead to the formation of monohaloalkanes. For example, the reaction of ethene with hydrogen chloride forms chloroethane. Similarly, the reaction of 2-butene with hydrogen bromide gives 2-bromobutane. Now you can see that the products of halogenation of alkanes and hydrogen halides addition to alkenes belong to the same group – monohaloalkanes. However, in case of alkanes, they are formed by substitution reaction but from alkenes they are formed by its addition reaction. The substitution reactions are difficult to control at monohaloalkane stage but addition of hydrogen halides to alkenes will always give only one monohaloalkane as long as the carbon atoms joined by the double bond are symmetrically substituted.

ethene + H-Cl → chloroethane

trans-2-butene + H-Br → 2-bromobutane

In the above examples, alkenes are symmetrical; the atom or group present on both carbon atoms joined by double bond is the same. Each carbon has one hydrogen atom and one methyl group. If the alkenes are unsymmetrical meaning those having different atoms/groups on the two carbon atoms joined by double bond, two products are possible from addition reactions.

For example, 1-butene is an unsymmetrical molecule because one of the carbon atoms attached by double bond has two hydrogen atoms while the other carbon attached by double bond has one hydrogen atom and one ethyl group. In this case, two addition products 1-bromobutane and 2-bromobutane are possible. So, one must know which one is the major product, and which one is the minor, and also what factor governs the formation of main product.

1-butene + H-Br → 1-bromobutnae + 2-bromobutane (main product)

The main addition product in reactions of unsymmetrical alkenes with unsymmetrical reagents is governed by Markovnikov's rule. According to this rule, the hydrogen atom (or the proton) goes to that carbon of the double bond which has the higher number of hydrogen atoms. That means that the remaining part of the reagent attaches to the carbon with the lower number of hydrogen atoms. Thus, in the reaction of 1-butene with hydrogen bromide the proton goes to 1st carbon, which has two hydrogen atoms and bromine (bromide) goes to 2nd carbon, which has only one hydrogen atom.

In the structure shown below, the carbons attached by double bond have one hydrogen atom each, but the alkyl groups attached to them are not same. One carbon has an ethyl group and other carbon atom has a propyl group. In this case, the halogen addition will give mixture of products.

+ H-Cl ⟶ 2-chloroheptane + 3-chloroheptane

Problem: 2-chlorobutane can be obtained either by addition of HCl to 2-butene or by substitution of *n*-butane using chlorine in the presence of light. Which method will you prefer and why?

Solution: The preferred method is the addition of HCl to 2-butene because the substitution of *n*-butane may lead to the formation of other products besides 2-chlorobutane.

Mechanism of hydrogen halides addition

The addition of hydrogen halides to alkenes takes place by an *ionic mechanism*. The hydrogen halides are ionized to a proton and a halide ion. The pi-electrons, because they are loosely bound as mentioned earlier in discussion of hybridization, capture the proton to form a carbon-hydrogen single bond leading to an intermediate ion. Since the other carbon attached to double bond has lost an electron that it was sharing, it will now have a positive charge in the intermediate. Such intermediates which bear a positively charged carbon are called *carbocation*. The negative halide ion will then react with this intermediate at the carbon atom bearing the positive charge to form a carbon-halogen bond. The steps involved in addition of hydrogen halide to ethene are shown below. You can easily see that the reaction is proceeding through ionic mechanism. Keep in mind that for the addition of unsymmetrical reagents to unsymmetrical alkenes, the proton goes to the carbon that has higher number of hydrogen atoms.

$$H{-}X \longrightarrow H^+ + X^-$$

$$H_2C{=}CH_2 + H^+ \longrightarrow H_3C{-}\overset{\oplus}{C}H_2 \xrightarrow{X^-} H_3C{-}CH_2X$$

Explanation of Markovnikov's rule

You have seen in the mechanism of addition of halogen halides that the reaction proceeds through an intermediate called carbocation. In the reaction of ethene and other symmetrically substituted alkenes, there will be only one possible carbocation. In unsymmetrical alkenes such as propene, however, there are two possible carbocations. The major addition product of the reaction is formed through the most stable carbocation. The carbocations are classified as primary (1°), secondary (2°), and tertiary (3°) depending on whether the positively charged carbon is attached to two, one or no hydrogen atoms. It has been established from many studies that the stability order of carbocations is as follows: tertiary (3°) > secondary (2°) > primary (1°) (**Figure 4.3**).

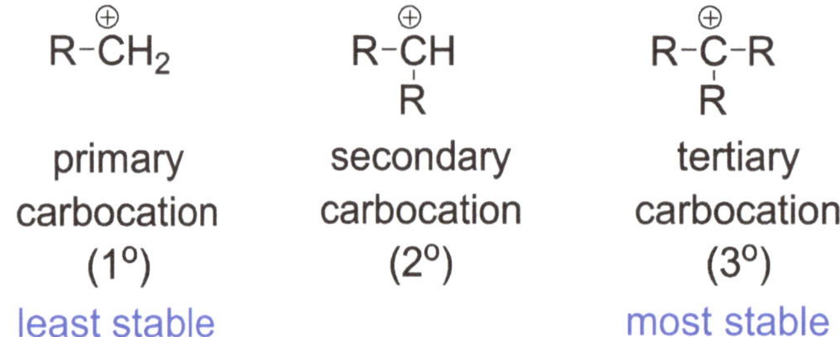

Fig. 4.3: *Primary, secondary, and tertiary carbocations.*

In the reaction of propene with H-X, if the proton attaches to terminal carbon, a secondary carbocation is formed. But if the proton is attached to the middle carbon, then a primary carbocation will be formed. Thus, the former carbocation is more stable. So, the main product in the reaction of propene with H-X, say HCl, is 2-chloropropane.

$H_3C-CH=CH_2$ $\xrightarrow{H-Cl}$ $H_3C-\overset{\oplus}{C}H-CH_3$ + $H_3C-CH_2-\overset{\oplus}{C}H_2$ (two possible carbocations)

(2° carbocation) more stable; (1° carbocation)

$\downarrow + Cl^{\ominus}$

$H_3C-CHCl-CH_3$

(2-chloropropane)

Addition of Water (Hydration): Preparation of Alcohols

The addition of water to alkenes requires an acid as a catalyst. The catalyst is necessary because water molecules do not ionize with the same ease as do hydrogen halides. An alkene and acid are mixed first, and then the mixture is poured slowly into water. The commonly used acids are sulfuric acid and phosphoric acid.

Markovnikov's rule applies in case of unsymmetrical alkenes. The product in this case too is not a hydrocarbon since it contains oxygen in the form of OH or hydroxyl group. The product belongs to the class of alcohols. The catalytic reaction of ethene with water produces ethanol or ethyl alcohol. It is used as antiseptic in hospitals, and in the production of alcoholic beverages. Thus, the reaction of alkenes with water is a method for preparation of alcohols. Alcohols are further used for synthesis of aldehydes, ketones, and acids by oxidation reactions.

$H_2C=CH_2$ + H-OH $\xrightarrow{H_2SO_4}$ H_3C-CH_2-OH

ethene → ethanol

1-butene + H-OH $\xrightarrow{H_2SO_4}$ 2-butanol (markonikov product)

Addition of alcohols: Preparation of ethers

The reaction of alkenes with alcohols in the presence of an acid as a catalyst leads to the formation of ether. For example, the reaction of methanol with propene in the presence of sulfuric acid forms isopropyl methyl ether as the major product according to the Markovnikov's rule. Guess what the minor ether product of this reaction will be.

propene + CH_3-OH $\xrightarrow{H_2SO_4}$ isopropyl methyl ether (markonikov product)

The mechanism of addition of water or alcohol to alkenes involves protonation of C=C bond followed by nucleophilic reaction of water with carbocation forming hydronium ion. The elimination of proton from the latter yields the final product.

c. Oxymercuration-Demercuration Reaction

There are some other important reactions of alkenes, in which alkenes are reacted with reagents other than water but the final product remains alcohol due to addition of H-OH. One of these reactions is oxymercuration-demercuration of alkenes which give alcohol in accordance with the Markovnikov's rule.

1. $Hg(OCOCH_3)_2$, H_2O
2. $NaBH_4$

d. Hydroboration-Oxidation Reaction

Another important reaction of alkenes that gives alcohol is hydroboration-oxidation sequence. The addition of borane to alkene, referred as hydroboration, forms alkyl borane which is oxidized by hydrogen peroxide to form alcohol. This reaction gives anti-Markovnikov's alcohol.

hydroboration

$$3\ H_3C{-}\underset{H}{C}{=}CH_2 + BH_3 \longrightarrow H_3CH_2CH_2C{-}B(CH_2CH_2CH_3)_2$$

borane tri-*n*-propylborane

oxidation

$$(CH_3CH_2CH_2)_3B + 3\ H_2O_2 + 3\ NaOH \longrightarrow 3\ CH_3CH_2CH_2OH + Na_3BO_3 + 3\ H_2O$$

The first step in this sequence is addition of a B-H bond to C=C. The B-H bond polarizes with a fractional positive charge on boron and a fractional negative charge on hydrogen. The positively charged boron adds onto the less substituted carbon. Thus, it resembles a normal electrophilic addition to an alkene in accordance with the Markovnikov's rule. However, the addition occurs in a concerted manner (all bond-breaking and bond-forming occurs in a single step).

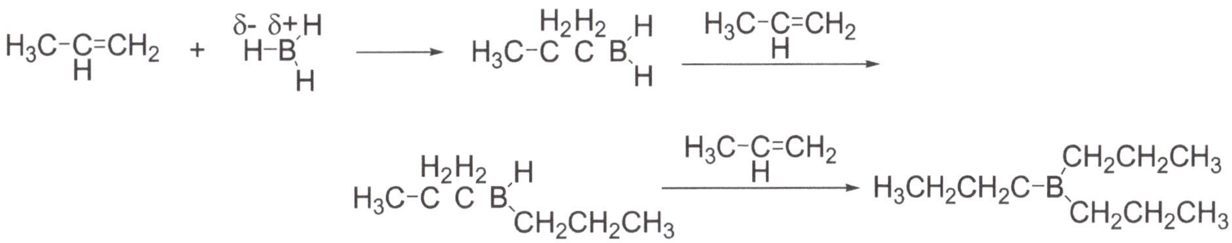

$$H_3C{-}\underset{H}{C}{=}CH_2 + \overset{\delta-\ \delta+}{H{-}B}H_2 \longrightarrow H_3C{-}\overset{H_2}{C}{-}\overset{H_2}{C}{-}BH_2 \xrightarrow{H_3C-\underset{H}{C}=CH_2}$$

$$H_3C{-}\overset{H_2}{C}{-}\overset{H_2}{C}{-}B(H)CH_2CH_2CH_3 \xrightarrow{H_3C-\underset{H}{C}=CH_2} H_3CH_2CH_2C{-}B(CH_2CH_2CH_3)_2$$

Problem: Write chemical equation showing the formation of following alcohols from vinylcyclohexane.

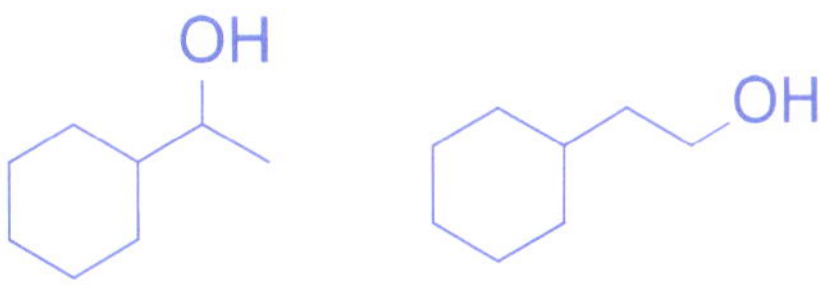

OH OH

Solution:

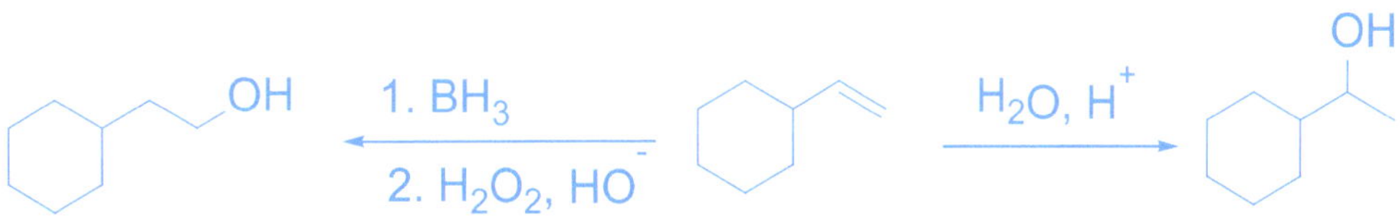

OH ← 1. BH_3 / 2. H_2O_2, HO^- — vinylcyclohexane — H_2O, H^+ → OH

e. Oxidation of Alkenes: Preparation of 1,2-Diols, Cyclic Ethers, and Carbonyl Compounds

Reactions with potassium permanganate

The reaction of alkenes with alkaline $KMnO_4$ gives *cis*-1,2-diols. 1,2-Diols are compounds containing two hydroxyl groups on adjacent carbon atoms. This reaction is also known as hydroxylation reaction. This reaction is used in laboratory as a chemical test for alkenes. As the reaction proceeds, the purple color of the permanganate ion is replaced by the brown precipitate of the manganese ion.

3 + 2 $KMnO_4$ + 4 H_2O —KOH→ 3 HO OH + $2MnO_2$ + 2 KOH

1,2-diol (glycol)

Oxidation of alkenes with $KMnO_4$ in acidic medium rather than in alkaline medium leads to the cleavage of the carbon-carbon double bond producing two carbonyl compounds (if alkene is symmetrically substituted only one carbonyl compound in double the molar ratio of the alkene). If a hydrogen atom is attached to the C=C, then the product from that side of the alkene will be a carboxylic acid. If any of the carbon of C=C has two hydrogen atoms, then one carboxylic acid is formed, and carbon dioxide is released.

+ $KMnO_4$ —H_3O^+→ O + O

3-methyl-2-butanone 2-hexanone

+ $KMnO_4$ —H_3O^+→ OH O + O OH

+ $KMnO_4$ —H_3O^+→ OH O + CO_2

Ozonolysis of alkenes

Alkenes react easily and rapidly with ozone producing ozonides. Ozonide can be explosive on isolation, so it is directly reacted with a reducing agent usually zinc and aqueous acid to give two carbonyl products as the isolated products (only one product if the alkene is symmetrically substituted as in the scheme shown below). However, if the double bond is present in a ring a dicarbonyl compound will be obtained as a single product.

O_3 → an ozonide → Zn, H_3O^+ → 2 (two carbonyls (only one product))

O_3 → Zn, H_3O^+ → two carbonyl compounds

1. O_3; 2. Zn-H_3O^+

Formation of epoxides

Alkenes react with various reagents to form three-membered cyclic ethers called epoxides or oxiranes. The most important commercial epoxide, ethylene oxide, is obtained by the silver-catalyzed air oxidation of ethylene. Other epoxides are usually prepared by the reaction of an alkene with an organic peroxyacid simply called as peracid.

$$H_2C{=}CH_2 + O_2 \xrightarrow[250^\circ,\ \text{pressure}]{\text{silver catalyst}} H_2C\overset{O}{-}CH_2$$

$$+ \quad R{-}\overset{O}{\overset{\|}{C}}{-}O{-}O{-}H \longrightarrow \quad + \quad R{-}\overset{O}{\overset{\|}{C}}{-}OH$$

f. Polymerization of Alkenes: Free-Radical Addition

Many simple alkene molecules undergo rapid addition reaction among themselves when treated with a small amount of a radical catalyst (R-O-O-R). One such catalyst is the benzoyl peroxide. This reaction is called polymerization and the product is called polymer (poly = many, mer = parts). When ethylene polymerizes, the product is called polyethylene. The polymerization of esters forms polyesters, used for fabric manufacturing.

$$H_2C{=}CH_2 \xrightarrow[1000\ \text{atm},\ >100\ ^{o}C]{R\text{-}O\text{-}O\text{-}R} *\!\!\left(\begin{matrix} H & H \\ | & | \\ C & - & C \\ | & | \\ H & H \end{matrix}\right)_{\!n}\!\!*$$

polyethylene

The homolytic cleavage (each oxygen atom takes one electron of the O-O single bond) of organic peroxide generates radicals. A catalyst radical then adds onto the C=C forming another carbon radical. This radical adds onto another alkene molecule and process continues until termination of the reaction chain, probably by combination of two radicals (dimerization of radicals).

$$R\text{-}O\text{-}O\text{-}R \xrightarrow{\text{homolytic cleavage}} 2\ R{-}\dot{O}$$

initiator radical

$$R{-}\dot{O} + H_2C{=}CH_2 \longrightarrow R{-}O{-}\overset{H_2}{C}{-}\dot{C}H_2 \xrightarrow{H_2C=CH_2} R{-}O{-}\overset{H_2}{C}{-}\overset{H_2}{C}{-}\dot{C}H_2$$

a carbon radical

and so on-----

Polymers are, thus, large molecules formed from the repeating units of a small molecule (monomer = single unit of the molecule). They have very high molecular weight. Synthetic polymers are perhaps the most important synthetic products that affect our day-to-day life. From what we wear (nylon, polyester) to what we stand on (floors with PVC (polyvinyl chloride) carpet) are synthetic polymers. There are number of well-known natural polymers as well. Silk, wool, and natural rubber are in use since centuries. Cellulose, a carbohydrate is the main constituent of wood and paper. Many more natural polymers are vital to life such as polysaccharides (carbohydrates), polypeptides (proteins), and nucleotides.

4.5 Chemical Properties of Alkynes

Alkynes burn like other hydrocarbons. Their flames are smoky due to the increased proportion of carbon atoms to hydrogen atoms in comparison to alkanes and alkenes. Alkynes are highly reactive compounds because of the presence of pi-electrons and undergo addition reactions like alkenes. They incorporate two equivalents of reagents instead of one equivalent in the case of alkenes. Some examples are given below.

a. Addition of Hydrogen

Alkynes are converted into alkanes using two mole equivalents of hydrogen in the presence of platinum or palladium or nickel as a catalyst.

$$\text{3-hexyne} + 2\,H_2 \xrightarrow{\text{Ni or Pt}} \text{hexane}$$

It is possible to obtain an alkene from hydrogenation of an alkyne using a specific catalyst called Lindlar's catalyst. The Lindlar's catalyst is prepared by precipitating palladium metal on calcium carbonate support and then deactivated by treatment with lead acetate and quinolone (an amine). The reaction of an alkyne in the presence of Lindlar's catalyst gives *cis* isomer of the product alkene.

$$\text{2-butyne} \xrightarrow[\text{Lindlar catalyst}]{H_2} \textit{cis}\text{-2-butene}$$

If one wishes to synthesize a *trans*-alkene from an alkyne by controlled hydrogenation, another set of reagents needs to be used. The reaction of alkenes with sodium in liquid ammonia affords a *trans*-alkene.

$$\text{2-butyne} \xrightarrow[\text{liquid } NH_3]{Na} \textit{trans}\text{-2-butene}$$

b. Addition of Halogens

Each molecule of an alkyne reacts with two halogen molecules and hence four halogen atoms are added to give tetrahaloalkanes. For example, one mole of 2-butyne reacts with two moles of chlorine to furnish 2,2,3,3-tetrachlorobutane.

$2\ Cl_2$

Cl Cl

Cl Cl

2-butyne

2,2,3,3-tetrachlorobutane

c. Addition of Hydrogen halides

This represents an addition of an unsymmetrical reagent. Since one alkyne molecule can accommodate two hydrogen halide molecules, the final product of the reaction will contain two halogen atoms and two more hydrogen atoms in comparison to the substrate alkyne. The final product is thus a dihaloalkane. Addition of the first mole of hydrogen halide to carbon-carbon triple bond will form a monohaloalkene. In this type of alkene, the carbon-carbon double bond would be unsymmetrically substituted. Hence it would react with another molecule of hydrogen halide according to the Markovnikov's rule forming the dihaloalkane. However, if the alkyne itself is unsymmetrical it would react with first molecule of hydrogen halide also according to the Markovnikov's rule.

2-butyne

2 HCl

HCl

HCl

H_3C H CH_3 H Cl Cl

2,2-dichlorobutane

$H_3C-C=C-CH_3$ (H, Cl)

2-chlorobutene

d. Addition of Water

Addition reaction of water with alkynes gives ketone as the final product. This reaction requires a mercury(II) salt such as mercury(II) acetate as a catalyst besides an acid. The immediate product of hydration is called enol (because it contains *ene* and *ol*). This product can exist in two isomeric forms. The more stable form is ketone (contains C=O group). The two isomers are called tautomers. It is a special type of isomerism in which two forms of the products are in dynamic equilibrium, with the more stable product dominating in the equilibrium mixture.

2-butyne + H_2O $\xrightarrow[HgSO_4]{H_2SO_4}$ enol ⇌ 2-butanone (a ketone)

Problem: How can you synthesize 2-hexanone from 1-hexyne.

Solution: 2-Hexanone can be synthesized from 1-hexyne by its hydration.

$H_3C-CH_2-CH_2-CH_2-C\equiv CH \xrightarrow[H^+,\ Hg^{2+}]{H_2O}$ 2-hexanone

4.6 Acidity of acetylenic Hydrogen: Applications in Organic Synthesis

The hydrogen atom attached to a triply bonded carbon (sp-hybridized) is weakly acidic. This hydrogen atom can be removed as a proton by a strong base such as sodium amide in liquid ammonia solution to form a carbanion. The carbanion being nucleophilic can react with several electrophiles. One simple example is its reaction with organic halides. Since the carbon-halogen bond is a highly polar bond, the carbon atom attached to halogen in halides bears a fractional positive charge. The alkyne carbanion reacts at this carbon and expels the halogen atom as halide ion. Thus, the whole sequence of reactions of alkynes with sodium amide and organic halides results into an increase of carbon chain length of alkyne. For example, ethyne is changed to propyne by the reactions shown below. The second hydrogen atom attached to acetylenic carbon can be removed in similar way and any other electrophile can be reacted with the resulting carbanion.

$$\underset{\text{ethyne}}{H-C\equiv C-H} \xrightarrow[\text{liq. } NH_3]{Na^+NH_2^-} \underset{\text{sodium acetylide}}{H-C\equiv \overset{\ominus}{C}\,Na^+} + NH_3$$

$$H{-}C{\equiv}C^{\ominus}\,Na^{+} \xrightarrow{H_3C{-}Br} H{-}C{\equiv}C{-}CH_3 + NaBr$$

propyne

Problem: Show a reaction scheme for the preparation of *cis*-2-hexene from propyne.

Solution:

$$H_3C{-}C{\equiv}CH \xrightarrow[\text{liq. } NH_3]{NaNH_2} H_3C{-}C{\equiv}C^{\ominus}Na^{\oplus} \xrightarrow{Br{-}CH_2{-}CH_2{-}CH_3} H_3C{-}C{\equiv}C{-}CH_2CH_2CH_3$$

$$\downarrow \text{Lindlar's catalyst}$$

$$H_3C(H)C{=}C(H)CH_2CH_2CH_3 \ (cis)$$

SUMMARY

Alkenes and alkynes are unsaturated aliphatic hydrocarbons. In naming branched alkenes or alkynes, the chain containing double or triple bond is considered as parent chain even if it is not the longest chain. If a compound has both carbon-carbon double- and triple bond, and both are equidistant then the double bond gets priority.

Many alkenes exhibit geometrical isomerism. Isomers are called *cis-trans* isomers. The origin of geometrical isomerism in alkenes is the rigid nature of carbon-carbon double bond. If both the carbon atoms joined by the double bond have two different atoms/groups attached to them, then the identical groups on two carbon atoms may be on same or opposite side of the double bond leading to existence of *cis* and *trans* isomers, respectively. Complex geometrical isomers are named according to *E-Z* convention in which groups/atoms attached to C=C are prioritized based on atomic number.

Reactions of Alkenes

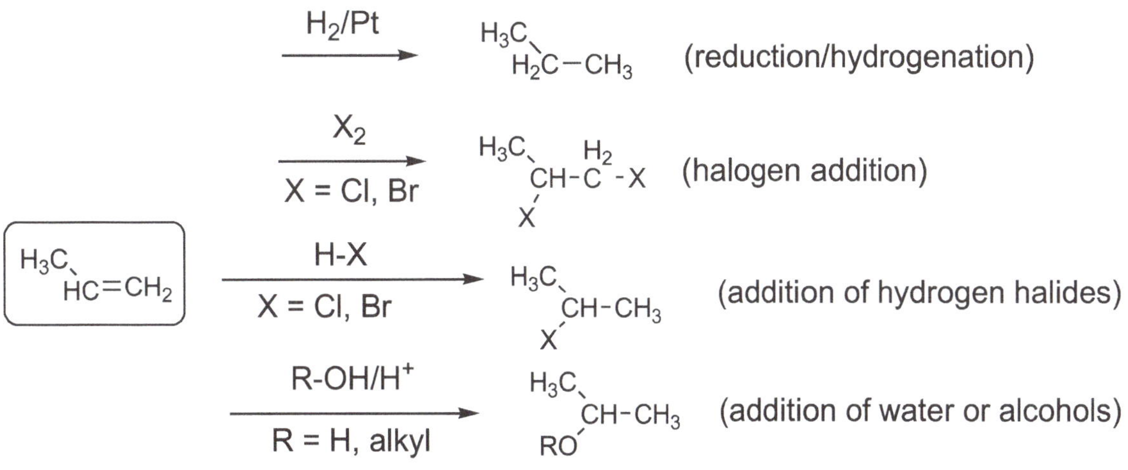

$H_3C-HC=CH_2$

$\xrightarrow{H_2/Pt}$ $H_3C-H_2C-CH_3$ (reduction/hydrogenation)

$\xrightarrow[X = Cl, Br]{X_2}$ $H_3C-CH(X)-CH_2-X$ (halogen addition)

$\xrightarrow[X = Cl, Br]{H\text{-}X}$ $H_3C-CH(X)-CH_3$ (addition of hydrogen halides)

$\xrightarrow[R = H, alkyl]{R\text{-}OH/H^+}$ $H_3C-CH(OR)-CH_3$ (addition of water or alcohols)

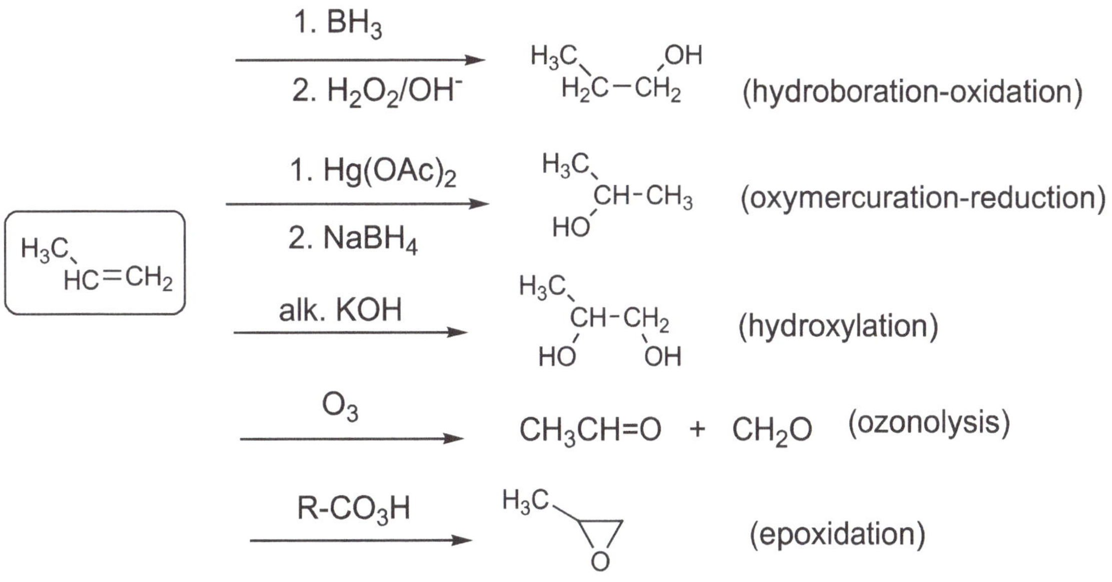

$H_3C-HC=CH_2$

$\xrightarrow[2.\ H_2O_2/OH^-]{1.\ BH_3}$ $H_3C-H_2C-CH_2-OH$ (hydroboration-oxidation)

$\xrightarrow[2.\ NaBH_4]{1.\ Hg(OAc)_2}$ $H_3C-CH(OH)-CH_3$ (oxymercuration-reduction)

$\xrightarrow{alk.\ KOH}$ $H_3C-CH(OH)-CH_2-OH$ (hydroxylation)

$\xrightarrow{O_3}$ $CH_3CH=O + CH_2O$ (ozonolysis)

$\xrightarrow{R\text{-}CO_3H}$ H_3C–(epoxide, O) (epoxidation)

Reactions of Alkynes

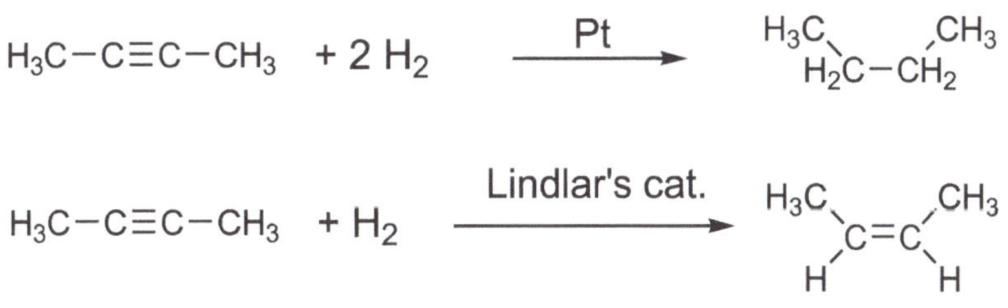

$H_3C-C\equiv C-CH_3 + 2\ H_2 \xrightarrow{Pt} H_3C-H_2C-CH_2-CH_3$

$H_3C-C\equiv C-CH_3 + H_2 \xrightarrow{Lindlar's\ cat.} H_3C(H)C=C(H)CH_3$

$$H_3C-C\equiv C-CH_3 \xrightarrow{Na/liq.\ NH_3} H_3C(H)C=C(H)CH_3$$

$$\xrightarrow[X = Cl, Br]{X_2}$$

$$\xrightarrow[X = Cl, Br]{H\text{-}X}$$

$$H_3C-C\equiv C-CH_3 \xrightarrow[H^+, Hg_2^+]{H_2O}$$

$$H_3C-C\equiv CH \xrightarrow[liq.\ NH_3]{NaNH_2} [H_3C-C\equiv C^{\ominus}] \xrightarrow{R\text{-}Br} H_3C-C\equiv C-R$$

EXERCISES

1. Write the structural formulas of following alkenes.

   a. 2-chloro-2-methyl-1-butene, 2. *trans*-3-hexene, c. *cis*-2-methyl-2-butene, d. 1,3-dimethylcyclopentene, e. *cis*-3-bromo-4-chlorocyclobut-1-ene, f. 1,4-dibromobut-2-ene, g. 1,4-butadiene, h. allyl chloride

2. Name the following hydrocarbons by the IUPAC system.

   a. $CH_3CH{=}C(CH_3)_2$, b. $CH_3CH{=}CHCH_2CH_3$, c. $CH_3C(Br){=}CHCH_3$,

   d. e. f. g. h.

   i. j. k. l.

3. Identify the following pairs of compounds as structural isomers, geometrical isomers and identical compounds.

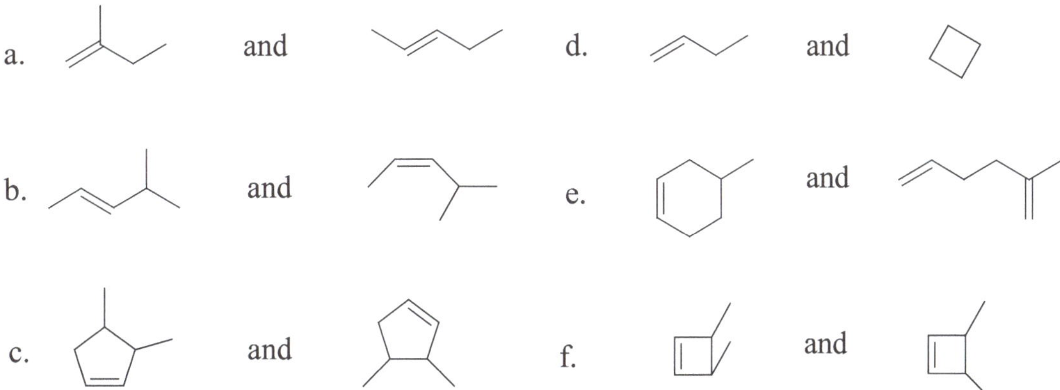

4. Which of the following compounds can exist as *cis-trans* isomers? Draw the structures of possible *cis-trans* isomers.

   a. 1-pentene, b. 1-chlorocyclohexene, c. 1-chloro-1-methylbutene d. 3-bromobut-1-ene, e. 2,3-dibromo-2-butene, f. 4,5-dimethylcyclohex-1-ene.

5. With the help of appropriate arrow, assign priority to groups/atoms attached to both carbon atoms of carbon-carbon double bond using *E-Z* convention. Identify the geometry at carbon-carbon double bond as *E* or *Z*.

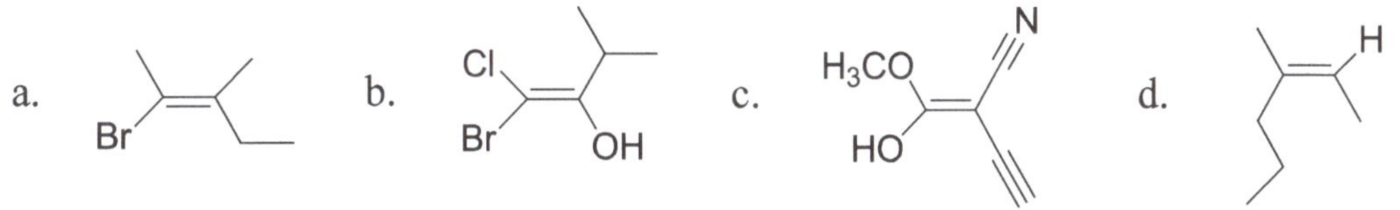

6. Briefly explain the following terms with suitable example(s).

   a. Hydroboration, b. Markovnikov's rule, c. polymerization

7. Write the structural formula and name of the product when one mole of each of the following compounds reacts with one mole of chlorine.

   a. $H_2C{=}CHBr$ b. c. d.

8. Complete the following reaction equations by providing missing alkene, reagent(s) or major product.

a. 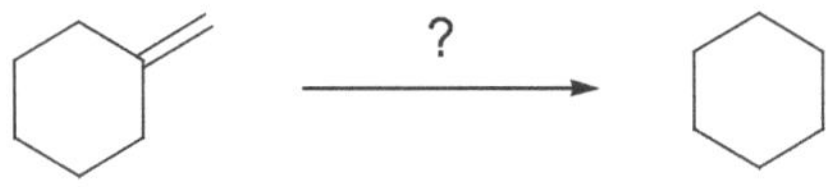 (methylenecyclohexane) —?→ (cyclohexane)

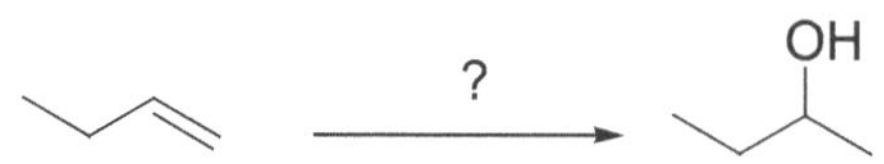

b. 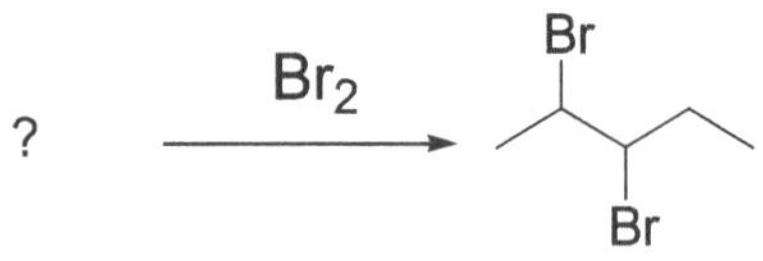 ? —Br_2→ (Br, Br dibromide)

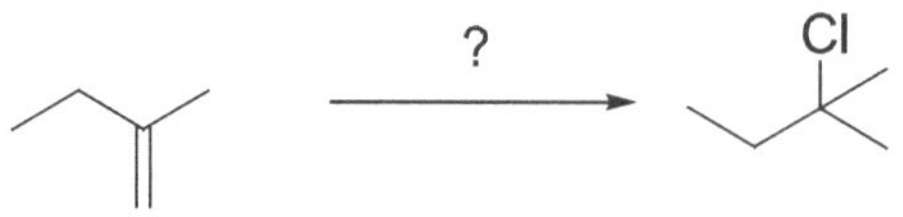

c.  (methylenecyclohexane) —HBr→ ?

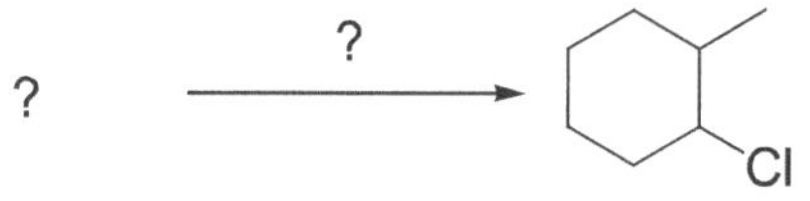

d.  ? —?→ (OCH_3 ether)

e. (alkene) —?→ (OH alcohol)

f. (alkene) —?→ (Cl chloride)

g. ? —?→ (Cl chloride)

h. 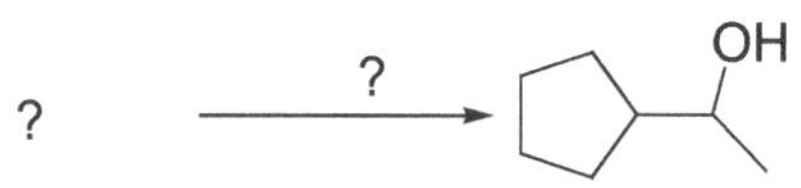 ? —?→ (OH alcohol)

9. Provide reagents to accomplish the conversion of following alkene to alcohols shown.

(methylenecyclohexane) ⟶ (OH alcohol) (methylenecyclohexane) ⟶ (OH alcohol)

10. The acid-catalyzed reaction of ethyl alcohol with 2-butene forms 2-butanol. Show stepwise mechanism of this reaction.

(2-butene) + C_2H_5-OH —H+→ (OC_2H_5 ether)

11. When methylpropene reacts with hydrochloric acid only one product 2-chloro-2-methylpropane is formed. Explain this finding showing the formation of possible carbocations and explaining their stability.

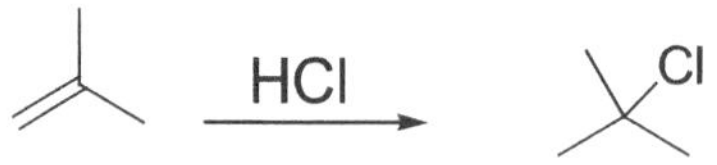

(methylpropene) —HCl→ (Cl chloride)

12. Write the product of the following reaction.

$+ \ KMnO_4 \ + \ H_2O \xrightarrow{KOH}$

$+ \ KMnO_4 \xrightarrow{H_3O^+}$

$+ \ KMnO_4 \xrightarrow{H_3O^+}$

$\xrightarrow[2.\ H_2O_2\text{-}OH^-]{1.\ BH_3}$

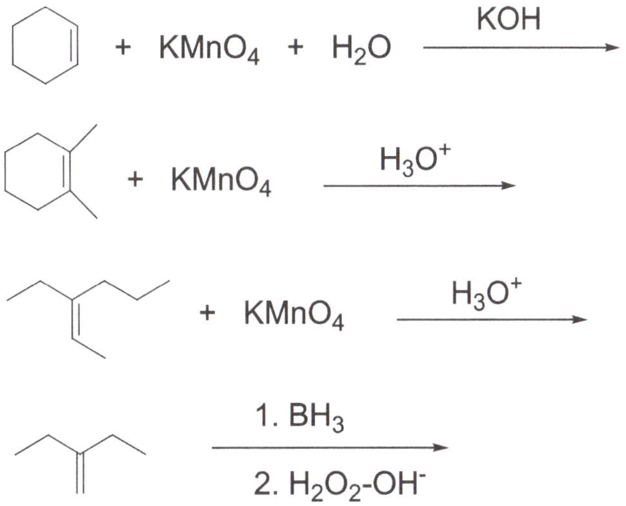

13. Ozonolysis of alkenes is known to cleave the C=C and give carbonyl compound(s). Give the structural formulas of alkenes that would give following carbonyl compound(s) on ozonolysis.

a. Two moles of CH_3CH_2CHO b. (CH3)$_2$CHO + CH_3CHO

c. 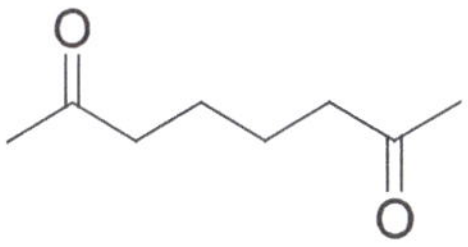 d.  +

14. Write equations for the following reactions.

a.2-butyne and excess hydrogen.

b. 2-butyne and 1 mole of hydrogen in the presence of Lindlar's catalyst.

c. 2-butyne and sodium in liquid ammonia.

d. 2-pentyne and two moles of chlorine.

e. 1-hexyne and two moles of HBr

f. 3-hexyne and water in the presence of H^+ and Hg^{2+} catalyst.

15. 2,3-Dibromopentane can be obtained from propyne in a four-step reaction sequence. Write the reaction scheme showing reagent and product of each step.

16. Explain with the help of chemical equations why 2-pentyne is able to react with two moles of HBr but only one mole of water.

17. Complete the following reaction scheme by providing the structure of missing reagent(s)/products.

$$? \xleftarrow[H^+,\ Hg^{2+}]{H_2O} HC{\equiv}CH \xrightarrow{?} HC{\equiv}C^{\ominus} \xrightarrow{C_2H_5Br} ?$$

$$HC{\equiv}CH \xrightarrow{?} H_2C{=}CH_2$$

$$? \xrightarrow{HBr\ (1\ mol)} ?$$

$$HOCH_2CH_2OH \xleftarrow{?} H_2C{=}CH_2 \xrightarrow{RCO_3H} ?$$

18. A hydrocarbon has molecular formula of C_3H_4 which corresponds to the molecular formula of propyne. However, one mole of this hydrocarbon reacts with only one mole of bromine while one mole of propyne is capable of reacting with two moles of bromine. Suggest the structure formula of hydrocarbon.

HYDROCARBONS-III: AROMATIC HYDROCARBONS

Aromatic hydrocarbons are highly unsaturated hydrocarbons with cyclic structure. The word "aroma" meaning flavor was used for such compounds that had typical fragrance such as benzene (obtained from coal distillate), toluene (obtained from *Tolu balsam*), and benzaldehyde (obtained from almonds, cherries, and peaches). Aromatic hydrocarbons are also called *arenes* due to the presence of carbon-carbon double bonds.

It was soon realized that such compounds had typical chemical property as well. For example, even though these compounds are highly unsaturated they don't show the typical chemical reactions of alkenes such as addition reactions under normal conditions (see reactions below). This is the reason why such compounds are studied as a separate class. The common examples of aromatic hydrocarbons are benzene, toluene, naphthalene, anthracene, and phenanthrene.

$KMNO_4$ / H_2O → OH, OH

H_2O-H^+ → H, OH

HBr-ether → H, Br

$KMNO_4$ / H_2O → no reaction

H_2O-H^+ → no reaction

HBr-ether → no reaction

The simplest aromatic hydrocarbon, benzene, has a six-membered carbon ring. Each carbon has a hydrogen atom giving it the molecular formula of C_6H_6. Thus, it has six hydrogen atoms less than its saturated analog cyclohexane (C_6H_{12}) indicating the presence of three carbon-carbon double bonds (C=C) in it.

Structural formula of benzene Line formula of benzene

5.1 Structure and Stability of Benzene

a. Resonance Model of Benzene

Benzene is a resonance hybrid of two contributing structures proposed by Sir Kekule (**Figure 5.1**). As mentioned earlier, benzene does not undergo typical addition reactions of alkenes. All carbon atoms in benzene are sp^2-hybridized. Thus, each carbon atom has an unhybridized *p*-orbital containing one electron in it. These orbitals overlap with each other forming a ring of six π electrons. These π electrons are, thus, not associated individually with any particular carbon atom. These electrons are said to be *delocalized*. As a result, there *is neither any true carbon-carbon double bond nor any true carbon-carbon single bond* in benzene molecule. In fact, studies have shown that all carbon-carbon bond lengths in benzene are the same (1.38 Å). This bond length of 1.38 Å is in between the normal bond lengths of a carbon-carbon single bond and a carbon-carbon double bond.

It means that the structure of benzene shown above is not its true representation. Benzene is represented by a hexagon and a circle inside it which represents the delocalized π bonds or ring of π electrons. Delocalization of π electrons provides additional strength to the ring. It is thus very difficult to break an aromatic structure and so benzene does not undergo such reactions easily in which its aromatic ring structure is destroyed.

benzene

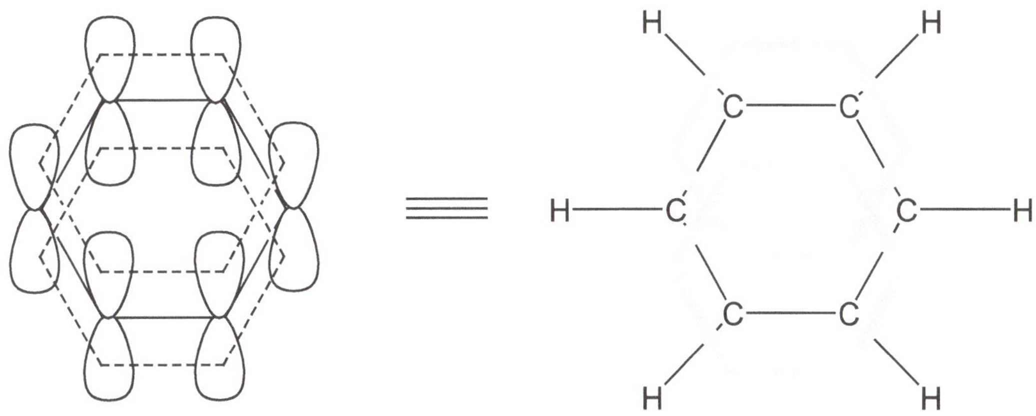

Fig. 5.1: *Molecular orbital picture of benzene.*

b. Stability of Benzene: Resonance Energy

You have studied in the preceding section that the delocalization of π electrons gives extra stability to the benzene ring. The increase in stability of the ring is attributed to the lower energy of the real molecule that is a resonance hybrid. The difference between the actual energy of the real molecule and the calculated energy of the most stable contributing structure is defined as the resonance energy. The experimental value of heat of hydrogenation of cyclohexene ring (one C=C) is 28.6 kcal/mole. Since the benzene ring has three carbon-carbon double bonds, the heat of hydrogenation, in theory, should be 28.6 X 3 kcal/mole that is 85.8 kcal/mol. The experimental value of the heat of hydrogenation of benzene, however, is only 49.8 kcal/mole. This value suggests that benzene ring is stabilized by 36 kcal/mol (85.8 kcal/mol – 49.8 kcal/mol). Thus, the resonance energy of benzene is 36 kcal/mol.

H_2/Pt → + 28.6 kcal/mole energy

H_2/Pt/ethanol, 2000 psi, 25 °C → + 49.8 kcal/mole energy

5.2 Nomenclature of Benzene Derivatives

Many benzene derivatives were known well before systematic nomenclature rules were developed. These compounds are popular with their common names (**Figures 5.2** and **5.3**) and are accepted by the IUPAC. So, you need to memorize these names. Some of the examples include:

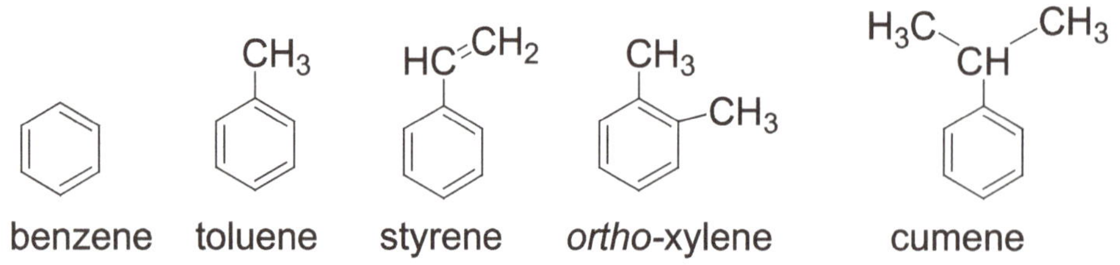
CH3
HC=CH2
CH3
CH3
H3C
CH
CH3
benzene
toluene
styrene
ortho-xylene
cumene

Fig. 5.2: *Common aromatic hydrocarbons.*

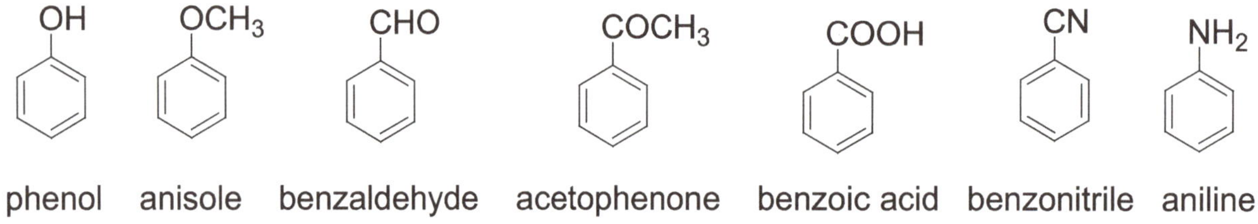
OH
OCH3
CHO
COCH3
COOH
CN
NH2
phenol
anisole
benzaldehyde
acetophenone
benzoic acid
benzonitrile
aniline

Fig. 5.3: *Common aromatic compounds.*

Those monosubstituted benzene compounds which do not have common names (**Figure 5.4**) accepted by the IUPAC are named as derivatives of benzene. For example,

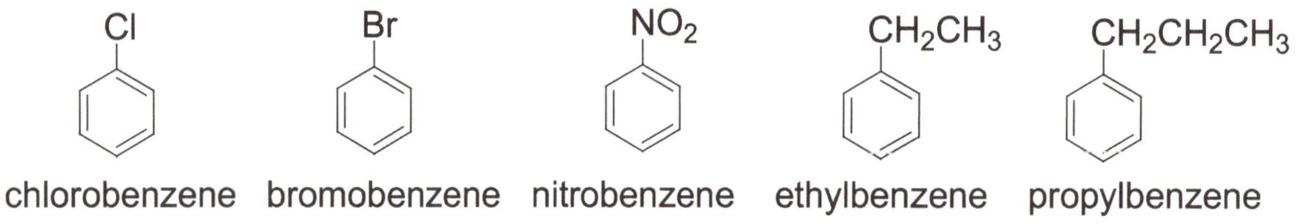
Cl
Br
NO2
CH2CH3
CH2CH2CH3
chlorobenzene
bromobenzene
nitrobenzene
ethylbenzene
propylbenzene

Fig. 5.4: *Benzene derivatives with no common names.*

If there are two substituents present on benzene ring, three isomers will be possible (**Figure 5.5**). In such cases, the position of substituents is located by using prefix *ortho* (2), *meta* (3) and *para* (4) which are abbreviated as *o*-, *m*-, and *p*-, respectively. Look at the two examples below. Benzoic acid has been treated as the parent molecule and the carbon atom bearing the COOH group has been numbered C-1. It has one bromo substituent which occupies *o*-, *m*-, and *p*-positions with regard to the COOH group.

COOH Br | COOH 1 2 3 Br 4 | COOH Br

ortho-bromobenzoic acid OR *o*-bromobenzoic acid OR 2-bromobenzoic acid

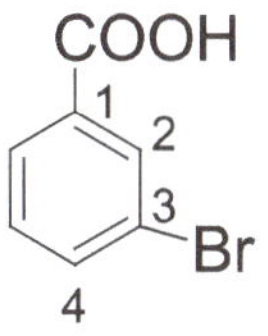

meta-bromobenzoic acid OR *m*-bromobenzoic acid OR 3-bromobenzoic acid

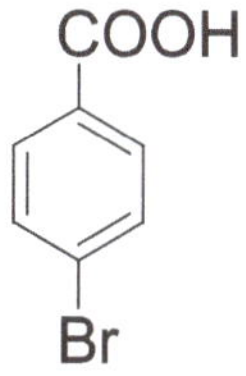

para-bromobenzoic acid OR *p*-bromobenzoic acid OR 4-bromobenzoic acid

Fig. 5.5: *Isomeric bromobenzoic acids.*

If the aromatic compound does not have a common name and two or more substituents are present on the benzene ring, the ring carbon atoms are numbered in such a way that the substituents get minimum number possible. In writing the name, substituents are arranged in alphabetical order.

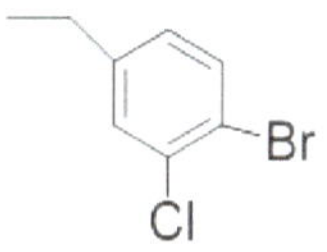

Br Cl

1-bromo-2-chloro-4-ethylbenzene

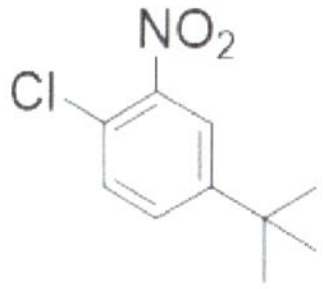

NO_2 Cl

4-(*tert*-butyl)-1-chloro-2-nitrobenzene

In certain compounds, benzene ring needs to be shown as a substituent (C_6H_5). In such cases it is called phenyl (abbreviated as Ph). The word "aryl" (abbreviated as Ar) is used in general for aromatic groups.

Ph_2CH_2

diphenylmethane

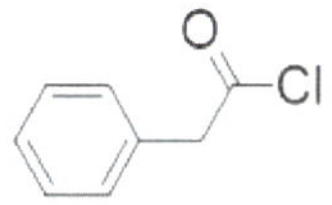

O Cl

2-phenylacetyl chloride

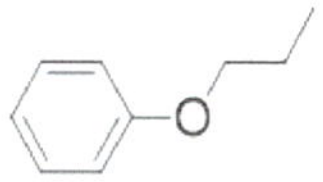

O

phenyl propyl ether

Problem: Draw the structure of *p*-bromophenol, 2-bromoacetophenone, and *m*-chlorobenzaldehyde.

Solution:

OH, Br — *p*-bromophenol; $COCH_3$, Br — 2-bromoacetophenone; CHO, Cl — *m*-chlorobenzaldehyde

5.3 Chemical Properties of Benzene

Benzene undergoes substitution reactions under the influence of some catalysts, in which aromatic ring remains intact. Some common products of substitution of benzene are alkyl benzenes, halobenzenes, nitrobenzene, benzenesulfonic acid and ketones (**Figure 5.6**). The substitution reactions leading to the formation of these products are called electrophilic aromatic substitution reactions. These benzene derivatives can be further transformed to several other compounds containing benzene rings. For example, the methyl group in methylbenzene is oxidized to get benzoic acid. Nitro group (NO_2) in nitrobenzene is reduced to amino group (NH_2) forming aniline.

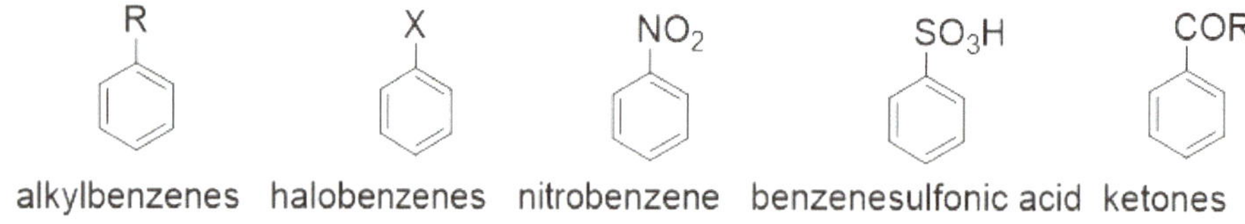

Fig. 5.6: *Some common benzene substitution products.*

a. Electrophilic Aromatic Substitution Reactions

Electrophilic aromatic substitution reactions of benzene open doors for the preparation of various other aromatic compounds. Benzene reacts with various electrophiles (electron-deficient species) resulting into the substitution of one or more hydrogen atoms of the benzene by the electrophile. Since the reactions involve substitution of an atom (hydrogen) present on the aromatic ring by an electrophilic species, the reactions are collectively referred as *electrophilic aromatic substitution reactions*. You must be clear in your mind that the benzene acts as a nucleophile (electron-rich species) in these reactions. Electrophilic aromatic substitution reactions of benzene include halogenation, nitration, sulfonation, Friedel-Crafts alkylation, and Friedel-

Crafts acylation. All these reactions, except the Friedel-Crafts alkylation, lead to the formation of products that are not hydrocarbons.

Halogenation of benzene: Formation of halobenzenes

The reaction of benzene with halogens such as chlorine or bromine in the presence of the Lewis acid catalysts such as $FeCl_3$ or $FeBr_3$ leads to the formation of halobenzenes such as chlorobenzene or bromobenzene. Remember that the reaction of alkenes with halogens was addition reaction and did not require any catalyst.

$$C_6H_6 + X_2 \xrightarrow{FeX_3} C_6H_5X + HX + FeX_3$$

X = Cl, Br

Nitration of benzene: Formation of nitrobenzene

The reaction of benzene with nitric acid in the presence of sulfuric acid as a catalyst leads to the nitration of benzene ring that is substitution of a ring hydrogen by a nitro group (NO_2) leading to the formation of nitrobenzene.

$$C_6H_6 + HNO_3 \xrightarrow[\text{(catalyst)}]{H_2SO_4} C_6H_5NO_2$$

Sulfonation of benzene: Formation of benzenesulfonic acid

Electrophilic aromatic substitution reaction of benzene with fuming Sulfuric acid (a mixture of SO_3 and H_2SO_4) results into sulfonation of the ring forming benzene sulfonic acid. This reaction is a reversible reaction.

$$C_6H_6 + \text{fuming } H_2SO_4 \rightleftharpoons C_6H_5SO_3H$$

Friedel-Crafts alkylation and acylation of benzene: Formation of alkyl benzenes and ketones

Alkylation and acylation reactions of benzene are commonly known as the Friedel-Craft's reactions after Charles Friedel and James Craft who discovered the reaction in 1877. The Friedel-Craft's alkylation is the reaction of benzene with alkyl halides, usually chloride or bromide in the presence of $AlCl_3$ or $AlBr_3$ as Lewis acid catalysts leading to the formation of alkyl benzenes. In a similar manner, benzene reacts with carboxylic acid chlorides or bromides to form ketones.

$$C_6H_6 + R\text{-}X \xrightarrow{AlX_3} C_6H_5R + HX + AlX_3$$

R = alkyl; X = Cl, Br

$$C_6H_6 + RCOX \xrightarrow{AlX_3} C_6H_5COR + HX + AlX_3$$

R = alkyl, aryl; X = Cl, Br

Problem: Write product of reaction between benzene and (i) propanoyl chloride (C_2H_5COCl) in the presence of aluminum chloride (ii) propene in the presence of an acid catalyst, and (iii) 1-chloropropane in the presence of aluminum chloride.

Solution: Benzene reacts with propanoyl chloride in the presence of aluminum chloride to form ethyl phenyl ketone. Propene and 1-chloropropane react with benzene through the most stable carbocation to form isopropyl benzene.

i) $C_6H_6 + CH_3CH_2COCl \xrightarrow{AlCl_3} C_6H_5COCH_2CH_3$

ii) $C_6H_6 + H_3C\text{-}CH{=}CH_2 \xrightarrow{H^+} C_6H_5CH(CH_3)CH_3$

iii) $C_6H_6 + CH_3CH_2CH_2Cl \xrightarrow{AlCl_3} C_6H_5CH(CH_3)CH_3$

b. Mechanism of Electrophilic Aromatic Substitution Reactions

The mechanisms of electrophilic aromatic substitution reactions can be described in three steps. The first step is the generation of electrophile (E^+) from the appropriate reagent. The second step is the reaction of benzene with electrophile forming *benzenonium ion*. Benzenonium ion is a carbocation, stabilized by resonance. The second step of the mechanism is also the slowest step of the reaction and hence the rate-determining step. The third and final step is aromatization of the ring by elimination of the proton and regeneration of the catalyst.

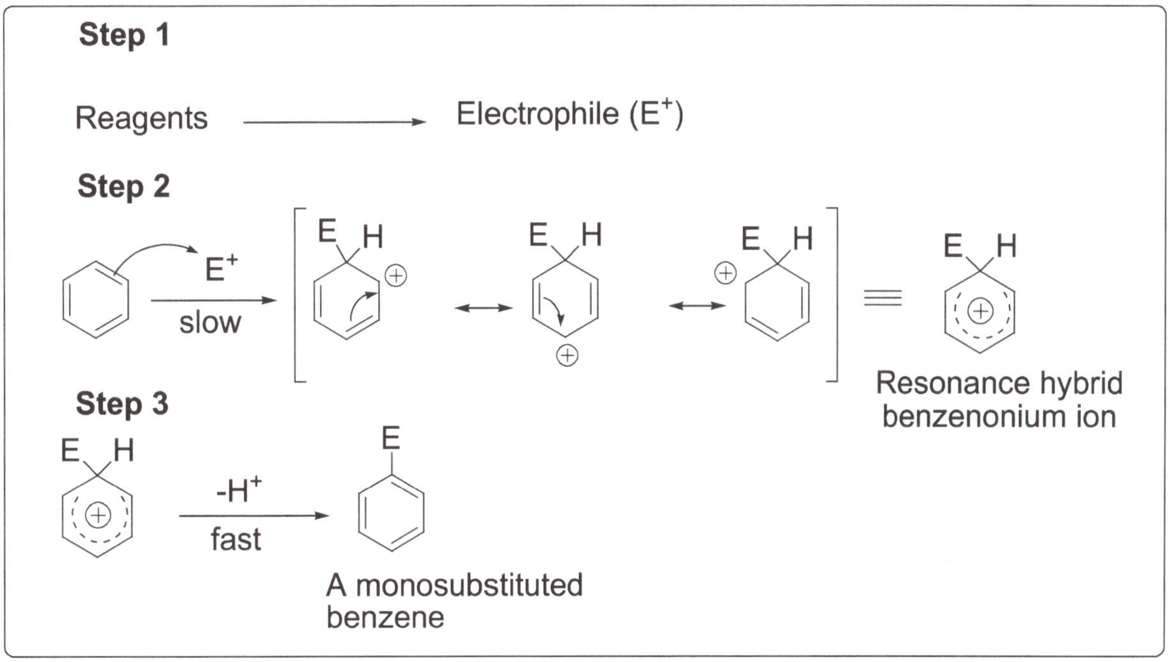

Step 1

Reagents ⟶ Electrophile (E^+)

Step 2

E^+, slow

Resonance hybrid benzenonium ion

Step 3

$-H^+$, fast

A monosubstituted benzene

After having an idea about the steps involved in electrophilic substitution of benzene ring, let's see the mechanisms of substitution reactions described in simplistic way.

Mechanism of chlorination

$$Cl_2 + FeCl_3 \rightleftharpoons Cl^+ + {}^-FeCl_4$$

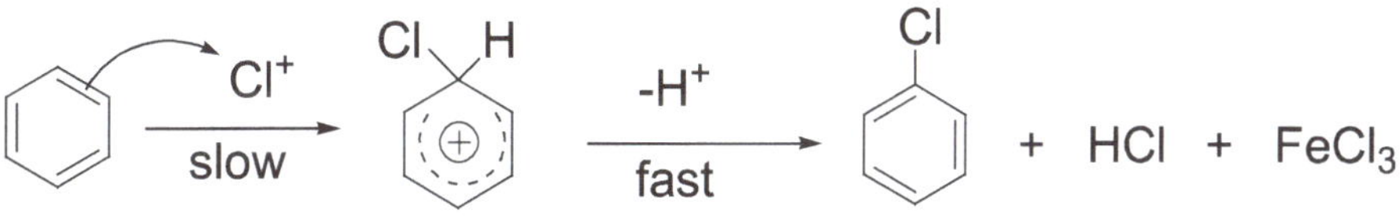

Cl^+, slow; $-H^+$, fast; $+ HCl + FeCl_3$

Mechanism of nitration

$$HO\text{-}NO_2 \underset{}{\overset{H^+ \text{(cat)}}{\rightleftharpoons}} {}^{\oplus}NO_2 + H_2O$$

$$\text{Benzene} + {}^{\oplus}NO_2 \xrightarrow{\text{slow}} \text{arenium ion } (O_2N,\ H) \xrightarrow[\text{fast}]{-H^+} \text{nitrobenzene } (NO_2) + H_3O^+$$

Mechanism of sulfonation

$$SO_3 + H\text{-}O\text{-}SO_2\text{-}OH \rightleftharpoons O{=}\overset{\oplus}{S}(=O)\text{-}OH$$

$$\text{Benzene} + {}^{\oplus}SO_3H \xrightarrow{\text{slow}} \text{arenium ion } (HO_3S,\ H) \xrightarrow[\text{fast}]{-H^+} \text{benzenesulfonic acid } (SO_3H)$$

Mechanism of Friedel-Crafts alkylation

$$CH_3CH_2\text{-}Cl + AlCl_3 \rightleftharpoons CH_3\overset{\oplus}{C}H_2 + \overset{\ominus}{A}lCl_4$$

$$\text{Benzene} \xrightarrow[\text{slow}]{CH_3\overset{\oplus}{C}H_2} \text{arenium ion } (H,\ CH_2CH_3) \xrightarrow[\text{fast}]{-H^+} \text{ethylbenzene } (CH_2CH_3) + HCl + AlCl_3$$

Mechanism of Friedel-Crafts acylation

$$H_3C\text{-}C(=O)\text{-}Cl + AlCl_3 \rightleftharpoons H_3C\text{-}\overset{\oplus}{C}{=}O + \overset{\ominus}{A}lCl_4$$

$$\text{Benzene} \xrightarrow[\text{slow}]{H_3C\text{-}\overset{\oplus}{C}{=}O} \text{arenium ion } (H_3COC,\ H) \xrightarrow[\text{fast}]{-H^+} \text{acetophenone } (COCH_3) + HCl + AlCl_3$$

Problem: When benzene reacts with propene or with 1-chloropropane the product is isopropylbenzene instead of *n*-propyl benzene. Explain.

Solution: It is because the reactions proceed through the intermediacy of more stable electrophile (carbocation).

$H_3C{-}CH{=}CH_2 \xrightarrow{H^+} H_3C{-}CH_2{-}\overset{\oplus}{C}H_2$ (less stable) $\quad H_3C{-}\overset{\oplus}{C}H{-}CH_3$ (more stable)

$H_3C{-}CH_2{-}CH_2{-}Cl \xrightarrow{AlCl_3} H_3C{-}CH_2{-}\overset{\oplus}{C}H_2$ (less stable) $\quad H_3C{-}\overset{\oplus}{C}H{-}CH_3$ (more stable)

5.4 Effect of Substituents on Reactivity of Benzene ring

The substituents already present on the benzene ring affect the reactivity of benzene ring towards electrophilic substitution reactions. Some substituents may increase the reactivity while some may decrease the reactivity. Those substituents which increase the reactivity are called ring-activating groups and those which diminish the reactivity are called ring-deactivating groups. Look at the nitration rate of following benzene derivatives taking the rate of nitration of benzene as standard (1). Phenol and toluene undergo nitration 1000 times and 24.5 times faster, respectively, than benzene itself. Chlorobenzene and nitrobenzene undergo nitration very slowly in comparison to benzene. Thus, hydroxyl group and methyl group are ring-activating groups whereas chloro and nitro groups are ring-deactivating groups.

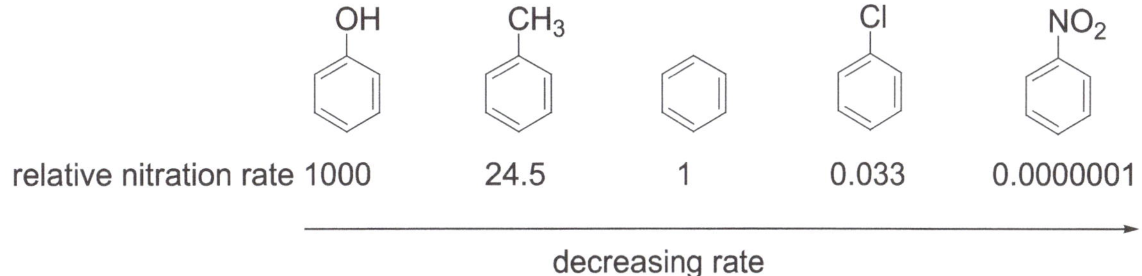

The substituents activate or deactivate the benzene ring by increasing or decreasing the electron density on the ring. They do so by inductive effect that depends on the electronegativity of the atom directly attached to the ring carbon and other atoms attached to this atom, and by resonance effect (see next section). The overall activating or deactivating effect and hence the reactivity of the benzene ring is determined by combined effect of these two factors. It is known from other evidences that hydroxyl and methyl groups are more electron-donating than hydrogen and hence ring-activating groups. Chloro and nitro groups are more electron-withdrawing than hydrogen and hence ring-deactivating groups. Among other important substituents or functional groups present on the benzene ring, alkoxy (e.g., OCH_3, OC_2H_5), and amino groups (NH_2, and $NHCH_3$) are ring-activating groups whereas carbonyl groups (e.g., COR, CO_2H, CO_2R, CHO), and nitrile (CN) are ring-deactivating groups.

5.5 Directing Effect of Substituents on Benzene Ring

A substituent already present on the benzene ring directs the position of another incoming substituent during the electrophilic substitution reactions of substituted benzenes. For example, when phenol is nitrated only two isomers of nitrophenol, 2-nitrophenol and 4-nitrophenol are obtained. 3-Nitrophenol is not obtained at all. When benzonitrile is nitrated, 3-nitrobenzonitrile is formed in 81% yield but 2-nitrobenzonitrile and 4-benzonitrile are formed only in a combined yield of 19%. Depending on the directing ability, the substituents on benzene ring are classified as *ortho-* and *para-*directing or *meta-*directing groups. The positions 2 and 4 from the substituent already present are called *ortho* and *para* positions, respectively; position 3 from the substituent is called *meta* position. Hydroxyl group is thus an *ortho-* and *para-*directing group whereas the nitrile group is a *meta-*directing group.

OH, HNO_3 / H_2SO_4, 25 °C → OH, NO_2 + OH, NO_2 + OH, NO_2

2-nitrophenol (*ortho*-nitrophenol) 50% | 4-nitrophenol (*para*-nitrophenol) 50% | 3-nitrophenol (*meta*-nitrophenol) 0%

CN, HNO_3 / H_2SO_4, 25 °C → CN, NO_2 + CN, NO_2 + CN, NO_2

2-nitrobenzonitrile (*ortho*-nitrobenzonitrile) 17% | 4-nitrobenzonitrile (*para*-nitrobenzonitrile) 2% | 3-nitrobenzonitrile (*mata*-nitrobenzonitrile) 81%

On detail and careful study, you will observe that all ring-activating substituents are *ortho*- and *para*-directing, and all ring-deactivating substituents except halogens are *meta*-directing groups. The halogens, even ring-deactivating, are *ortho*- and *para*-directing groups.

The directing effect of substituents can be explained by comparing the stability of different benzenonium ions, formed by reaction of a benzene derivative with an electrophile. For example, when toluene is nitrated three benzenonium ions can form due to reaction of nitronium ion at *ortho*-, *para*- and *meta*-positions (**Figure 5.7**). The benzenonium ions formed from reaction at *ortho*- and *para*-positions have contributions from a tertiary carbocation which makes the resonance hybrid more stable in comparison to resonance hybrid carbocation from *meta*-attack that has contributions only from secondary carbocations. As a result, the incoming substituent occupies mainly *ortho*- and *para*-positions.

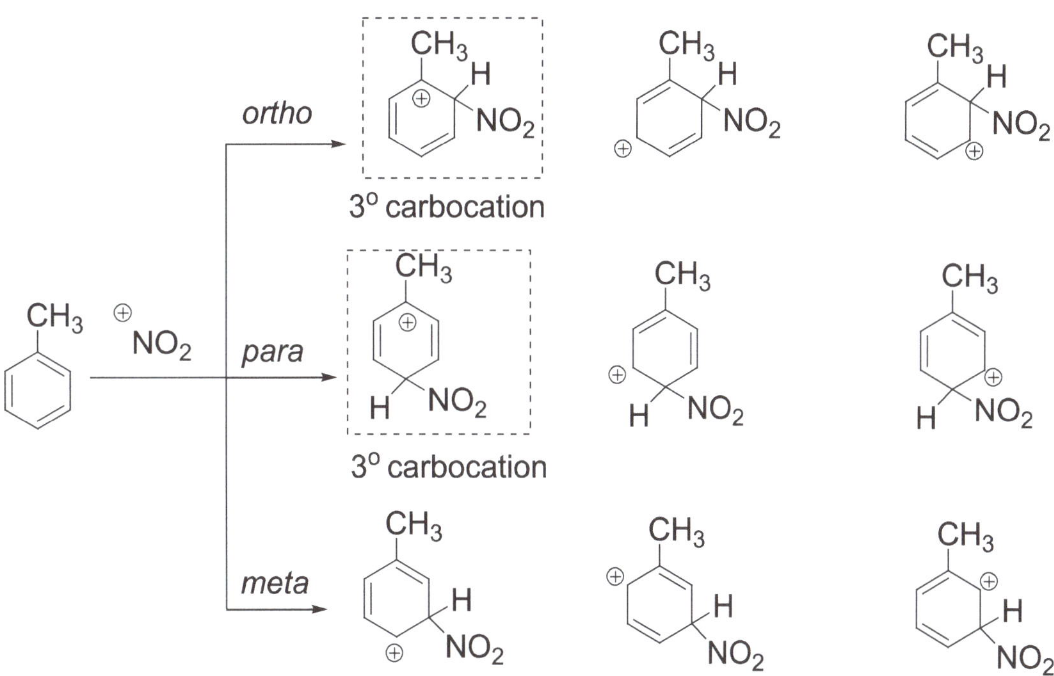

Fig. 5.7: *Nitration of toluene at meta-position explained.*

Phenols are even more reactive than toluene in nitration and give only *ortho*-nitrophenol and *para*-nitrophenol; no *meta*-nitrophenol at all. The tertiary carbocation formed in this case is further stabilized by resonance due to lone pairs of electrons on the oxygen atom of the hydroxyl group (**Figure 5.8**). Thus, the resonance-hybrid benzenonium ions obtained from reaction at *ortho*- and *para*-positions will be even more stable than those in case of toluene. Similar resonance stabilization of the benzenonium ions is possible in case of electrophilic substitution of halobenzenes also because of lone pairs on the halogen atom. Therefore, the chloro or bromo groups, even though ring-deactivating, are *ortho*- and *para*-directing groups.

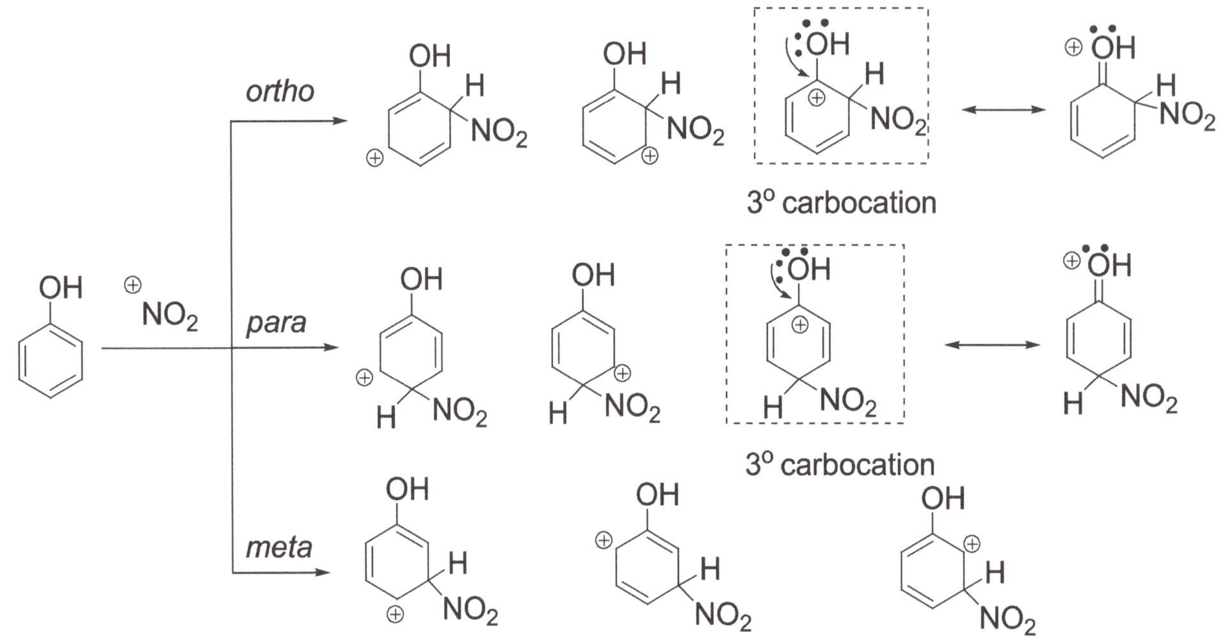

Fig. 5.8: *Nitration of phenol at ortho- and para-positions explained.*

Now let us explain the nitration of nitrobenzene which gives *meta*-dinitrobenzene predominantly. The nitrogen atom in nitro group has a formal charge of +1 (**Figure 5.9**). The one resonance contribution in benzenonium ions formed from reaction at *ortho* and *para* positions has a positive charge on carbon atom that is directly attached to nitrogen atom. This is a highly improbable contribution because the positive charges on adjacent atoms repel each other. There is no such structure possible from reaction at the *meta*-position. Therefore, the *meta*-substitution is preferred over *ortho*- and *para*-substitutions.

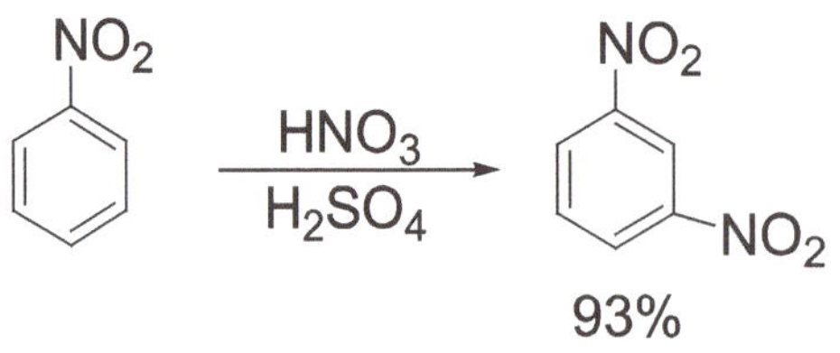

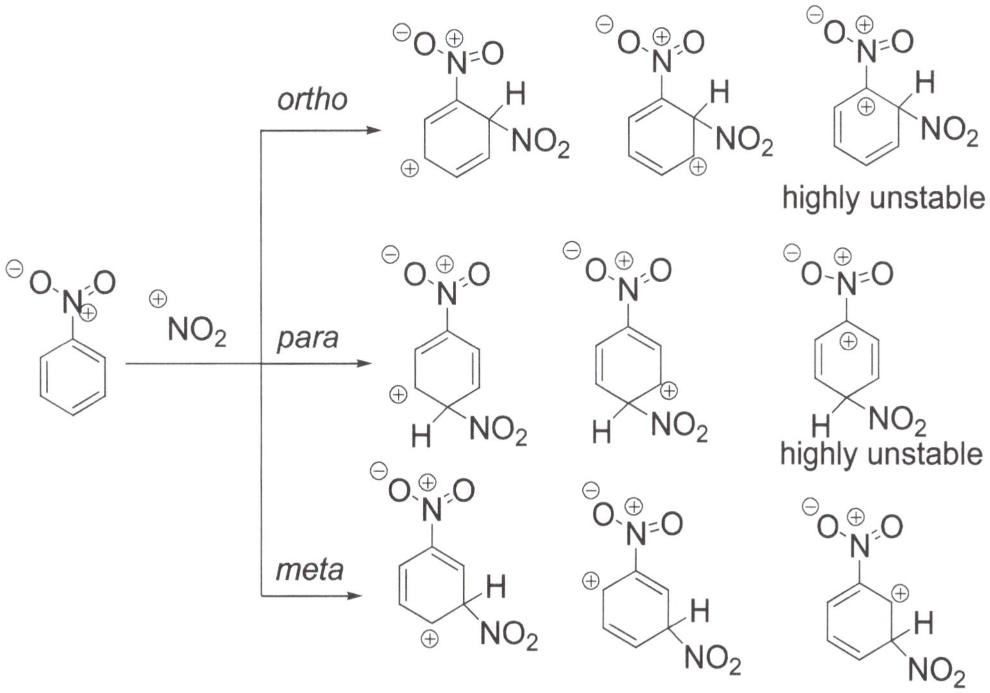

Fig. 5.9: *Nitration of nitrobenzene explained.*

In general, the alkyl groups and all other groups having unshared electron pair(s) on the atom directly attached to the benzene ring are *ortho*, *para*-directing groups. Further, all those groups in which the atom directly attached to the benzene ring is part of a multiple bond will be a *meta*-directing because the atom directly attached to the benzene ring will have a partial positive charge.

Problem: Write the products of the reaction between (i) methylbenzene (toluene) and bromine in the presence of $FeBr_3$, (ii) chlorobenzene and acetyl chloride in the presence of $AlCl_3$.

Solution:

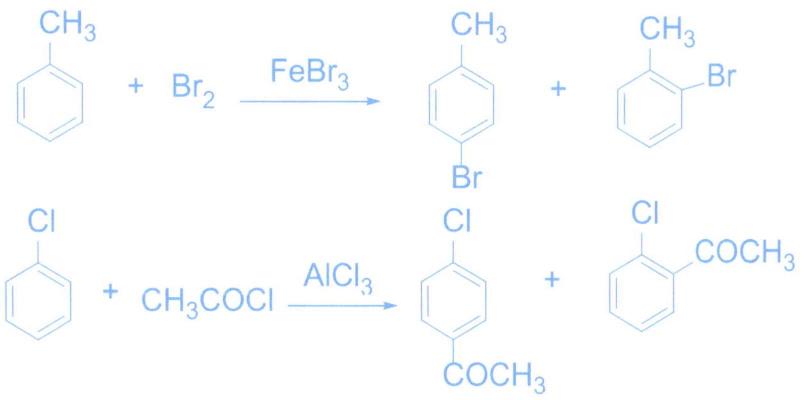

5.6 Significance of Directing Effect of Groups in Organic Synthesis

One needs to be careful in synthesizing a disubstituted benzene derivative due to different directing ability of the substituents. For example, if you want to synthesize *meta*-nitrotoluene from benzene, you must do nitration first to get nitrobenzene followed by methylation of the nitrobenzene (**Figure 5.10**). If you do methylation of benzene first, then the yield of *meta*-nitrobenzene would be very poor because the methyl group is an *ortho*-, *para*-directing group.

+ HNO_3 — H_2SO_4 (catalyst) → NO_2 — CH_3Cl, $AlCl_3$ → NO_2, CH_3

preferred method

CH_3Cl, $AlCl_3$ → CH_3 — HNO_3, H_2SO_4 (catalyst) → CH_3, NO_2

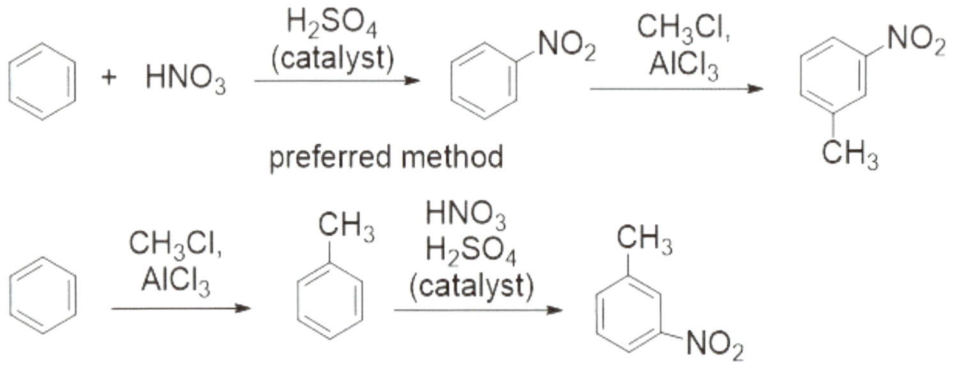

Fig. 5.10: *Synthesis of meta-nitrotoluene from benzene.*

Problem: Show the reaction scheme for synthesizing *para*-chlorotoluene from benzene.

Solution: You can start by synthesizing either chlorobenzene or tolucnc first because both chloro and methyl groups are *ortho*-, *para*- directing groups. However, synthesizing toluene is the better strategy because methyl group is an electron-donating group, so activates the benzene ring. The chloro group is a ring-deactivating group. Also, in both the steps, both *ortho*-, *para*-products will be formed that need to be separated.

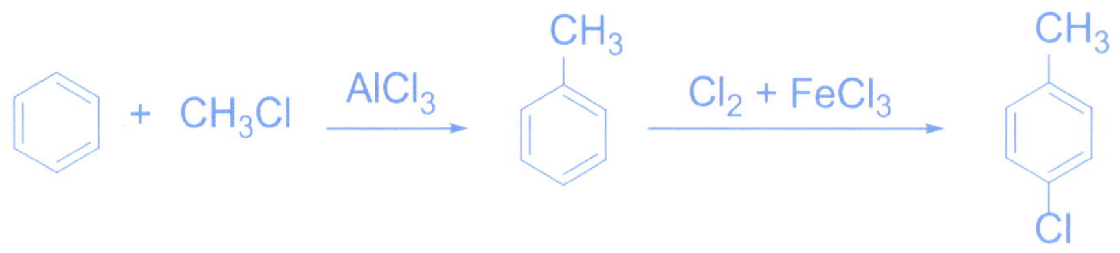

+ CH_3Cl — $AlCl_3$ → CH_3 — Cl_2 + $FeCl_3$ → CH_3, Cl

5.7 Application of Benzene Derivatives in Organic Synthesis

Benzene derivatives are very important for synthesis of other functional groups. For example, the nitro group of nitrobenzene can be reduced with tin in hydrochloric acid to amino group giving rise to aniline. Hydroxide base is added in the reaction to hydrolyze the amine hydrochloride to free amine. As another important example, the alkyl groups (having benzylic hydrogen) on benzene can be oxidized with potassium permanganate to carboxylic group forming benzoic acid.

NO_2 — Sn-HCl, $^{\ominus}OH$ → NH_2
aniline

CH_3 — aqueous $KMnO_4$ → COOH
benzoic acid

Problem: Show the reaction scheme for the synthesis of *meta*-nitrobenzoic acid from benzene.

Solution: The benzene is nitrated first and then treated with methyl chloride in the presence of aluminum chloride to get *meta*-nitrotoluene. The oxidation of methyl group in *meta*-nitrotoluene with aqueous potassium permanganate will give *meta*-nitrobenzoic acid.

+ HNO_3 —H_2SO_4→ NO_2 —CH_3Cl, $AlCl_3$→ NO_2, CH_3 —aqueous $KMnO_4$→ NO_2, COOH
meta-nitrobenzoic acid

5.8 Some Other Aromatic Compounds

All the compounds obtained in the reactions of benzene described above are aromatic compounds (not aromatic hydrocarbons only). All compounds in which a functional group is directly attached to the aromatic ring are called aromatic compounds. For example, phenol, benzoic acid, benzaldehyde, acetophenone, nitrobenzene, aniline, etc. are aromatic compounds but not aromatic hydrocarbons.

OH	COOH	CHO	$COCH_3$	NO_2	NH_2
phenol	benzoic acid	benzaldehyde	acetophenone	nitrobenzene	aniline

Naphthalene, anthracene, and phenanthrene are other simpler arenes (aromatic hydrocarbons). Naphthalene is a bicyclic compound that is used in moth balls whereas anthracene and phenanthrene are tricyclic compounds. It should, thus, be clear that arenes are not necessarily monocyclic. Pyridine is an example of heterocyclic (heterocyclic meaning ring which contains atom(s) other than carbon) aromatic compound (not aromatic hydrocarbon).

Polycyclic aromatic hydrocarbons and other aromatic compounds also show unusual reactivity like benzene. All these compounds, in general, have flat/planar structure in general and they obey Huckel's (4n + 2)π (where n = an integer with value1, 2, 3) electrons rule. Naphthalene has n =2, so 10 π electrons.

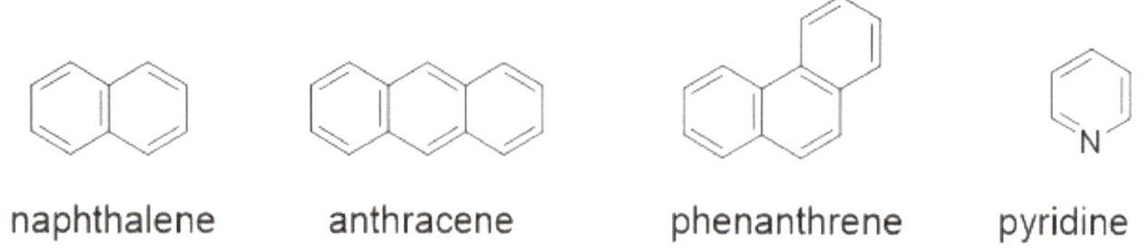

Problem: Count the number of π electrons in anthracene, phenanthrene and pyridine and find the corresponding integer values.

Solution: Naphthalene has 10 π electrons. Anthracene and phenanthrene have 14 π electrons. Pyridine has 6 π electrons. Thus, n is 2 for naphthalene, 3 for anthracene and phenanthrene, and 1 for pyridine.

SUMMARY

All planar cyclic hydrocarbons that contain 4n + 2 π-electrons are aromatic hydrocarbons. Benzene is the simplest possible aromatic hydrocarbon. Even though benzene has three carbon-carbon double bonds it shows unusual stability by resisting to addition reactions that are common to alkenes. This behavior of benzene is due to resonance-hybrid structure of benzene in which the π-electrons are delocalized. All carbon-carbon bonds in benzene have same bond length that is in between the length of a carbon-carbon single bond and carbon-carbon double bond. Thus, benzene does not have true carbon-carbon double bond.

Benzene and other aromatic hydrocarbons, under influence of certain catalysts, undergo substitution reactions with electrophiles. The most common electrophilic aromatic substitution reactions are halogenations, nitration, sulfonation, Friedel-Crafts alkylation and Friedel-Crafts acylation reactions. These reactions take place in three steps – (i) generation of electrophile, (ii) reaction of electrophile with benzene forming a resonance-stabilized benzenonium ion, and (iii) removal of proton leading to aromatization of the ring. The second step of the mechanism is rate-determining step.

$$+ \ HNO_3 \xrightarrow{H_2SO_4 \text{ (catalyst)}} NO_2$$

$$+ \ X_2 \xrightarrow{FeX_3} X \ + \ HX \ + \ FeX_3$$

X = Cl, Br

$$+ \text{ fuming } H_2SO_4 \rightleftharpoons SO_3H$$

$$+ \ R{-}X \xrightarrow{AlX_3} R \ + \ HX \ + \ AlX_3$$

R = alkyl; X = Cl, Br

$$+ \ R{-}C({=}O){-}X \xrightarrow{AlX_3} R \ + \ HX \ + \ AlX_3$$

R = alkyl, aryl; X = Cl, Br

Substituted benzenes, that may or may not be hydrocarbons, are important substrates in organic synthesis. For example, an alkyl group bearing a benzylic hydrogen on benzene can be oxidized by potassium permanganate to a carboxylic group forming benzoic acid. A nitro group on benzene ring can be reduced to amino group forming aniline. Further, these groups affect the reactivity of benzene ring and direct the position of incoming substituent if another substituent is to be introduced on a monosubstituted benzene ring. The presence of an electron-donating group on the benzene ring increases the reactivity of the ring towards electrophiles whereas the presence of an electron-withdrawing group on the benzene ring decreases the reactivity of the ring towards electrophiles. A substituent on benzene ring can be *ortho-*, *para*-directing or *meta*-directing. The groups with lone pair(s) of electrons on the atom directly attached to the benzene ring are *ortho-*, *para*-directing. However, if the atom attached to the benzene ring is involved in a double or triple bond formation the group will be *meta*-directing.

EXERCISES

1. 1. Write the structural formula of the following compounds.

   a. *m*-chlorotoluene b. *p*-bromobenzoic acidc. *m*-dinitrobenzene

   d. 4-chloroaniline e. 1,3-dimethylbenzenef. ethylbenzene

2. Write all possible structures of (i) trinitrotoluene, and (ii) tribromobenzene.

3. Draw all possible structures of an aromatic hydrocarbon with the molecular formula C_9H_{12} containing a benzene ring.

4. Give name to following aromatic compounds.

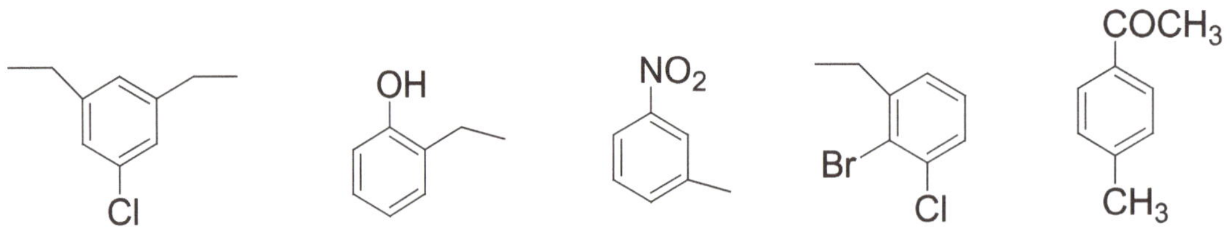

Cl

OH

NO_2

Br Cl

$COCH_3$ CH_3

5. Briefly explain why benzene does not undergo typical addition reactions with HCl or hydroxylation reaction with alkaline $KMnO_4$ similar to alkenes.

6. What simple chemical test you can use to distinguish benzene and styrene?

7. Briefly explain the terms: (a) aromaticity, and (b) resonance energy.

8. Complete the following chemical equations by providing the major product(s).

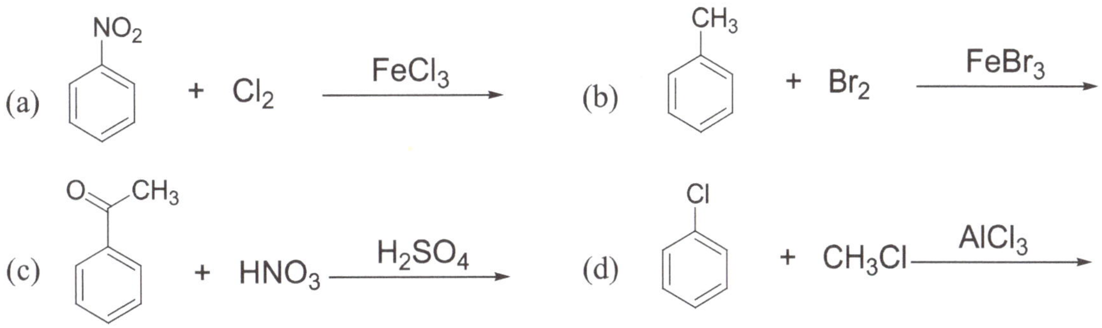

(a) NO_2 + Cl_2 $\xrightarrow{FeCl_3}$ (b) CH_3 + Br_2 $\xrightarrow{FeBr_3}$

(c) O CH_3 + HNO_3 $\xrightarrow{H_2SO_4}$ (d) Cl + CH_3Cl $\xrightarrow{AlCl_3}$

9. Write the mechanism of formation of products in the following reactions.

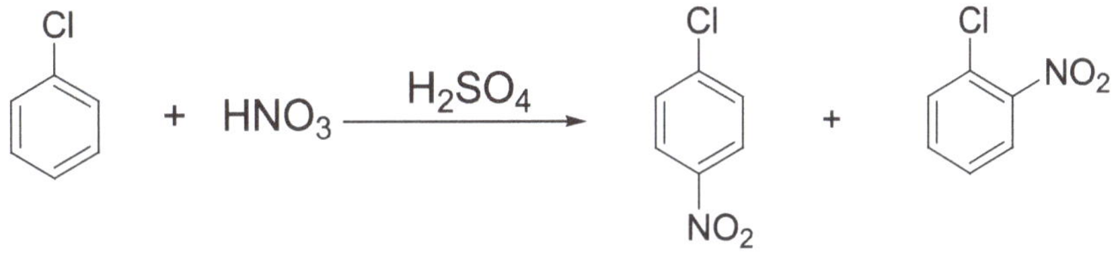

Cl + HNO_3 $\xrightarrow{H_2SO_4}$ Cl, NO_2 + Cl, NO_2

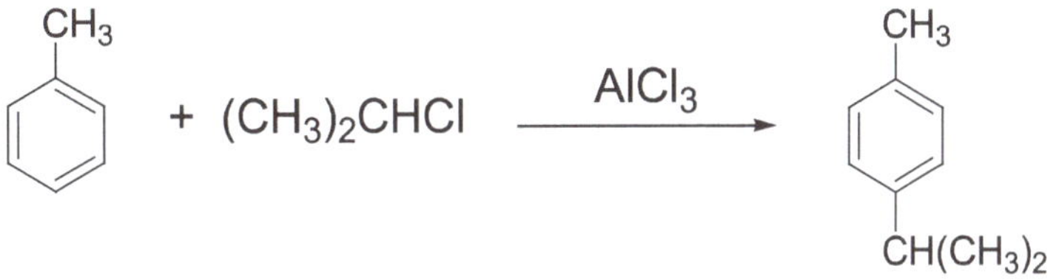

CH_3 + $(CH_3)_2CHCl$ $\xrightarrow{AlCl_3}$ CH_3, $CH(CH_3)_2$

10. Draw all possible contributing resonance structures to the benzenonium ions formed from the reaction of nitronium ion ($^+NO_2$) with chlorobenzene and explain why the major products are *ortho*- and *para*-substituted; not *meta*-substituted.

11. Identify the following groups as *ortho-*, *para*-directing or *meta*-directing.

a. $COOCH_3$, b. COOH, c. $CONHCH_3$, d. Br, e. OC_2H_5, f. $NHCH_3$, g. CHO

12. Which compound in the following pairs is more reactive towards electrophilic aromatic substitution and why?

a. Benzene and toluene b. chlorobenzene and toluene

b. Aniline and nitrobenzene d. benzoic acid and ethylbenzene

13. Show the stepwise reaction scheme mentioning all the reagents for the synthesis of following compounds starting from benzene.

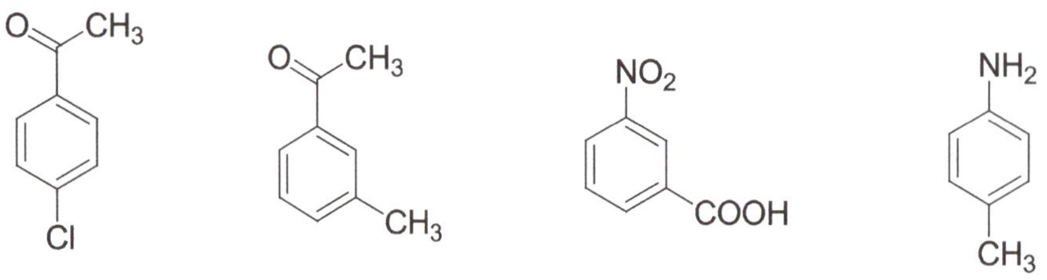

14. Cyclohexene reacts with bromine without any catalyst to form 1,2-dibromocyclohexane but benzene, even though having 6 π electrons does not react with bromine without catalyst. Explain.

15. When benzene is treated with 2-methylpropene in the presence of sulfuric acid, two substitution products are possible. Write structure of both possible products. Select the major product giving reason for your choice.

16. Keeping in mind the *ortho, para*-directing or *meta*-directing effect of the groups on benzene ring, write the structure of disubstituted benzene derivatives obtained in the following reactions.

(a) 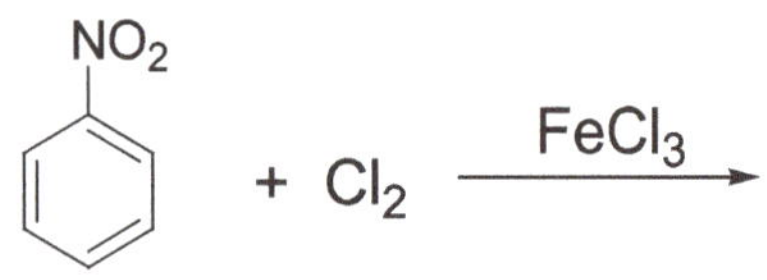 $C_6H_5NO_2$ + Cl_2 $\xrightarrow{FeCl_3}$

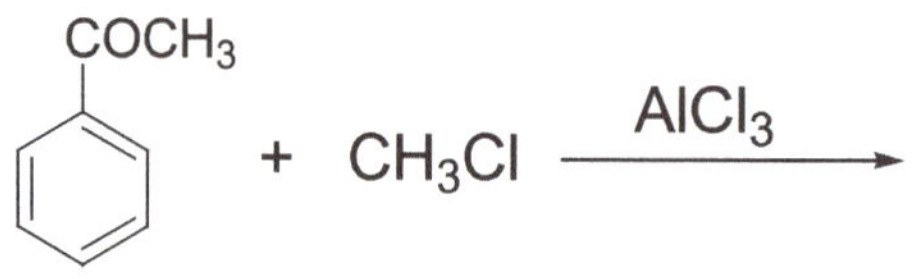

(b) 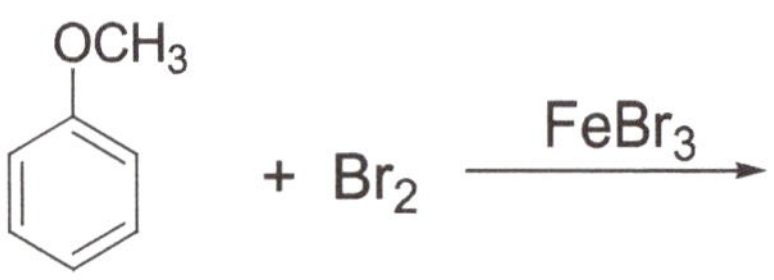 $C_6H_5OCH_3$ + Br_2 $\xrightarrow{FeBr_3}$

(c) $C_6H_5CH_3$ + HNO_3 $\xrightarrow{H_2SO_4}$

(d) $C_6H_5COCH_3$ + CH_3Cl $\xrightarrow{AlCl_3}$

(e) C_6H_5Br + CH_3COCl $\xrightarrow{AlCl_3}$

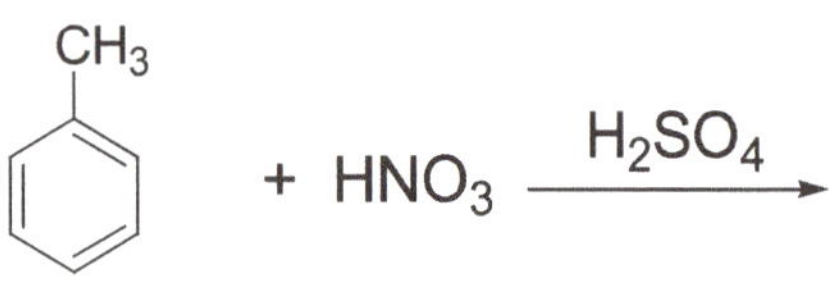

(f) 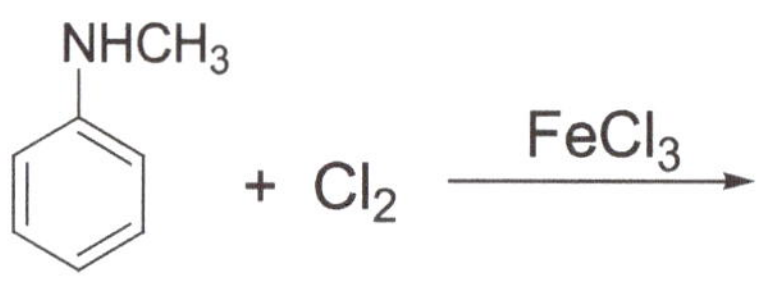 $C_6H_5NHCH_3$ + Cl_2 $\xrightarrow{FeCl_3}$

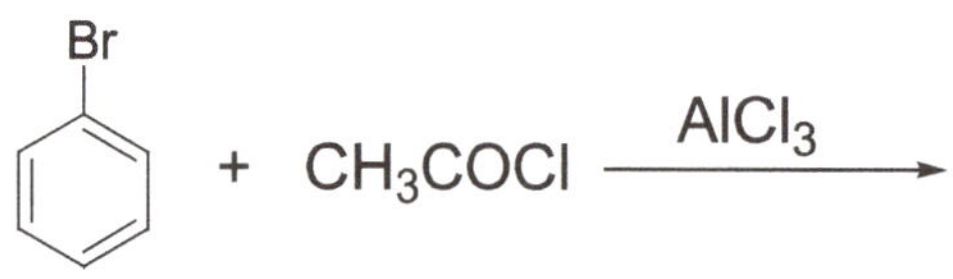

6 STEREOISOMERISM

The development of knowledge in the field of medicinal chemistry has made the study of stereoisomerism and stereochemistry an extremely important area of study for organic chemists. If there are two stereoisomers possible for a particular compound, one can be extremely active against a particular pathogen while other may not be active at all. Furthermore, the study of stereochemistry is also important to get an insight of the mechanism of organic reactions. You will see in the next chapter how the knowledge of stereochemical structure of the product helps us in determining the mechanism of reactions. Most of the compounds in our body such as carbohydrates, proteins, and enzymes occur in a particular stereoisomeric form. In this chapter, you will learn about optical activity of the compounds, chirality, and its origin in the molecule, diastereomers, and enantiomers.

6.1 Optical Activity and Plane-Polarized Light

Certain substances rotate the plane of the plane-polarized light passing through it. Such substances are called optically active compounds. The phenomenon was first discovered by Jean Baptiste Biot in 1813 who observed that when the plane-polarized light passed through a single crystal of quartz, or aqueous solution of sugar or tartaric acid, it was rotated either to left or right. Some other common examples are shown in the following figure (**Figure 6.1**).

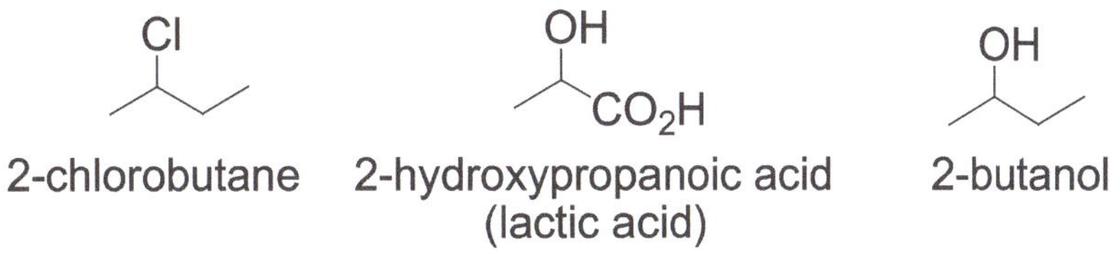

Cl
2-chlorobutane

OH
CO_2H
2-hydroxypropanoic acid
(lactic acid)

OH
2-butanol

Fig. 6.1: *Optically active molecules.*

Those compounds which rotate the plane toward right (clockwise) are called dextrorotatory (from the Latin *dexter* meaning "right") while those rotating the plane to the left (anticlockwise) are called levorotatory (from the Latin *laevus* meaning "left").

An ordinary light can be described as a wave that vibrates in all possible planes perpendicular to its path (**Figure 6.2**). The plane-polarized light vibrates in only one of these possible planes.

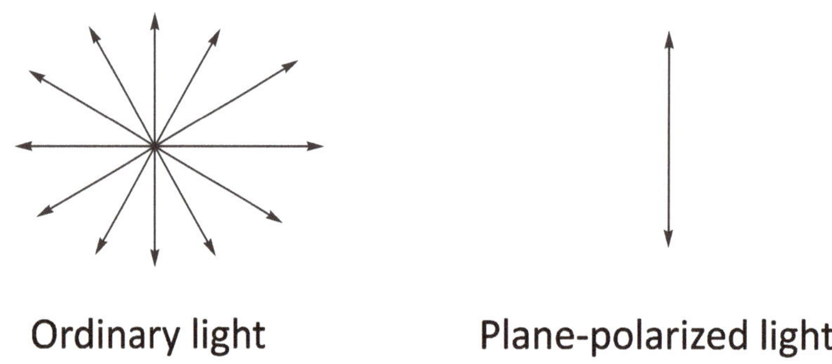

Fig. 6.2: *Ordinary and plane-polarized light.*

The device used to polarize the plane of the light was invented in 1828 by the British physicist William Nicol. It was composed of crystalline calcium carbonate and was called Nicol prism after the name of the discoverer. Later, another material called polaroid was developed by the American scientist E. H. Land. Polaroid contained a crystalline organic compound properly oriented and embedded in transparent plastic.

The optical activity is detected and measured by an instrument called polarimeter. A schematic representation of the setup is shown in figure (**Figure 6.3**). The compound in a sample tube is placed in between the path of the plane-polarized light and an analyzer. The observed angle of rotation is denoted by α. It depends on factors such as number of molecules in the sample tube, the length of the sample tube, the wavelength of the plane-polarized light, and temperature besides the molecular structure of the compound. In order to compare the optical activity of different substances, all these parameters are standardized by using the specific rotation $[a]_D$. The specific rotation of a compound is defined as the observed rotation when light of 589.6 nm wavelength is employed with a sample path length of 1 dm and sample concentration of 1 g/cm^3 (Equation 6.1).

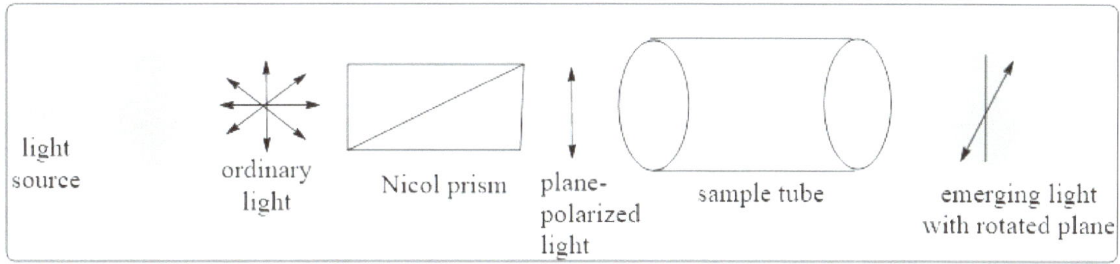

Fig. 6.3: *A schematic representation of use of polarimeter for measuring optical activity.*

$$specific\ rotation = [\propto]_D^t = \frac{\propto}{lc}$$ *Equation ---- (6.1)*

Where, α = observed rotation, t = temperature

l = length of the sample tube (1 dm = 10 cm), and

c = concentration of the sample in g/cm^3

6.2 Optical Isomers: Enantiomers and Chirality

Optical isomers are stereoisomers (**Figure 6.4**) that are identical in structure except where they differ in three-dimensional arrangement of atoms or groups at one or more carbon atoms and rotate the plane of the plane-polarized light differently.

CO_2H / H–C–OH / CH_3 | mirror | CO_2H / HO–C–H / CH_3

Fig. 6.4: *Mirror image: Optical isomers.*

Origin of Optical Activity: Chirality

Most compounds do not rotate the plane of the plane-polarized light. How is it that some do? It is not the particular class of organic compounds to which optical isomers belong to since they are found in all classes. So, what structural features in the molecule are responsible for optical activity of the compounds and existence of optical isomers?

Let's compare the structures of lactic acid (2-hydroxypropanoic acid) and propanoic acid (**Figure 6.5**). Lactic acid is optically active compound but propanoic acid is not.

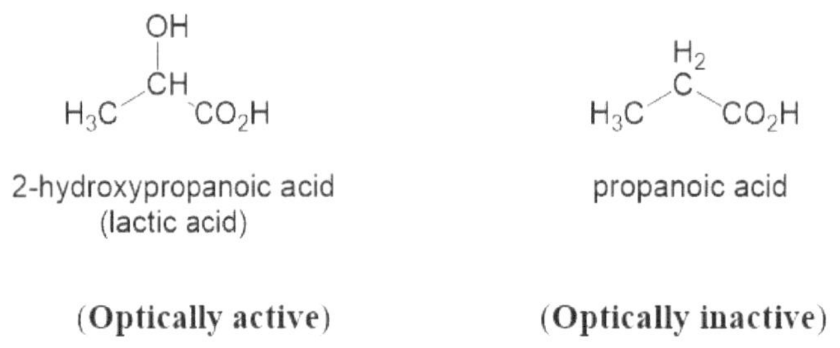

2-hydroxypropanoic acid (lactic acid)

(Optically active)

propanoic acid

(Optically inactive)

Fig. 6.5: *Structures of lactic acid and propanoic acid.*

In the structure of lactic acid, look at the second carbon atom (C-2). The four atoms/groups attached (H, CH_3, OH, CO_2H) to C-2 are different. There is no such carbon in the molecule of propanoic acid. Such a carbon atom in any molecule to which four different atom/groups are attached is called a chiral carbon. The presence of a chiral carbon in the lactic acid molecule provides chirality - handedness (from Greek meaning hand) to the molecule. In chiral compounds, the molecule and its mirror image are non-superimposable just like our left and right palms are non-superimposable mirror image of each other. Thus, two different arrangements of atom or groups are possible around the chiral carbon. These two different arrangements are related to each other as an object and its mirror image. These two mirror image structures or stereoisomers rotate the plane of the plane-polarized light equally but in opposite directions and hence are called optical isomers. Thus, optical isomerism exists due to chirality in the molecule. These optical isomers that are non-superimposable mirror images are called enantiomers. Thus, the mirror image optical isomers shown in Figure (**Figure 6.4**) above are enantiomers. See below other examples of 2-butanol and 2-chlorobutane (**Figure 6.6**).

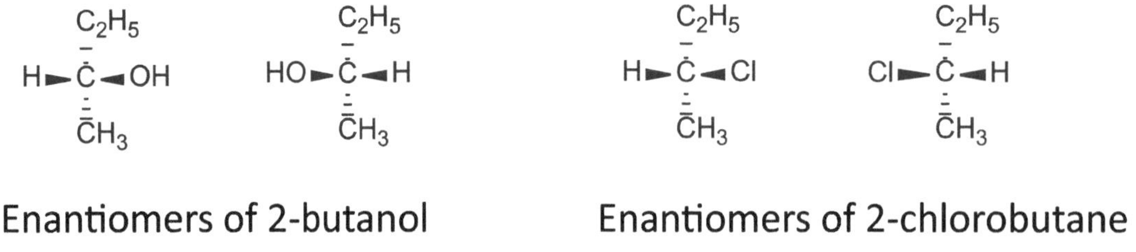

Enantiomers of 2-butanol

Enantiomers of 2-chlorobutane

Fig. 6.6: *Enantiomers of 2-butanol and 2-chlorobutane.*

Now let's have a look at the mirror image structures of propanoic acid (**Figure 6.7**). They are identical structures, same compound. Thus, the molecule of propanoic acid is *achiral* and hence optically inactive.

Fig. 6.7: *Propanoic acid: Achiral (optically inactive).*

Please keep in mind, however, that the presence of a chiral carbon is an important but not essential criterion for a compound to exhibit chirality. Chirality is the property of a molecule which depends on different kinds of *asymmetry* in the molecule. There can be other structural features besides the presence of a chiral carbon in the molecule that can lead to asymmetry or chirality in the molecule. On the other hand, you will find examples of compounds that contain more than one chiral carbon atoms, but they are achiral. This happens due to presence of certain symmetry elements in the molecule. For example, look at the stereoisomers of 2,3-dihydroxysuccinic acid (tartaric acid) (**Figure 6.8**). The first two structures are enantiomers. However, the rightmost structure has a plane of symmetry. The plane of symmetry divides the molecule into two identical halves. Such molecules are optically inactive because the optical rotation caused by the upper half of the molecule is cancelled by an equal rotation in opposite directions by the lower half of the molecule. This isomer is called *meso* isomer. The more appropriate term for the factor imparting chirality or asymmetry in the molecule, therefore, is stereogenic center. So, the chiral carbon can be called stereogenic carbon or stereogenic center as well.

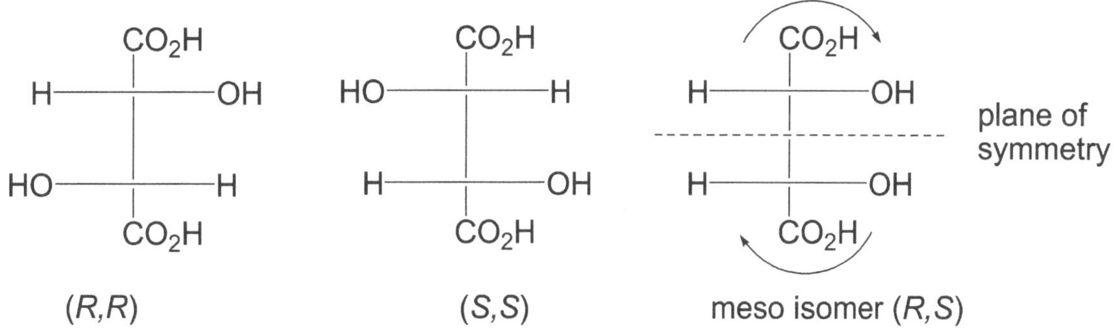

Fig. 6.8: *2,3-Dihydroxysuccinic acid: meso isomer.*

The molecules with more than one stereocenter can also exist as stereoisomers that are not mirror images or enantiomers. Look at the structures below. The top configurations in the molecules are opposite to each other but the configuration at the bottom stereocenter is same. The two molecules are, thus, not mirror image to each other. They are, however, stereoisomers. Such configurational stereoisomers that are not mirror images are called diastereomers (**Figure 6.9**).

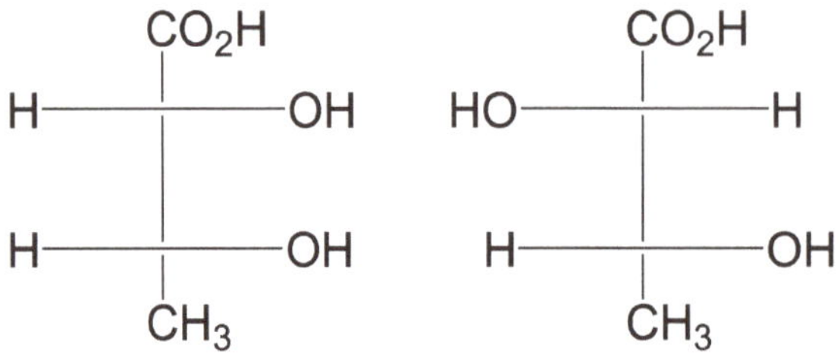

CO_2H CO_2H

H—|—OH   HO—|—H

H—|—OH   H—|—OH

CH_3 CH_3

Fig. 6.9: *Non-mirror image stereoisomers: diastereomers.*

Problem: Draw the structure of a chiral chloroalkane.

Answer: This should be a compound in which a carbon atom bearing a chlorine atom should be further attached to three different groups/atoms.

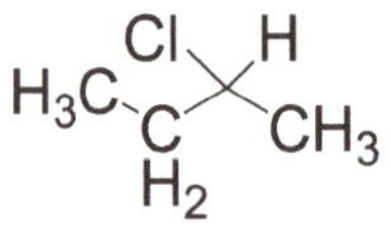

Cl H

H_3C–C(H_2)–C–CH_3 2-chlorobutane

Problem: Draw the structure of 4-methylheptane and tell whether it is chiral or achiral.

Solution:

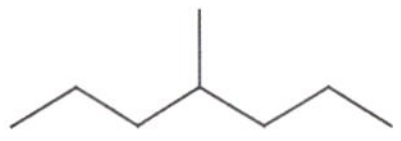

4-methylheptane

4-Methylheptane is achiral. There is no carbon atom attached to four different groups or atoms.

6.3 Racemic Mixture

A mixture of enantiomers in equal amounts is called racemic mixture or racemate. The racemate is optically inactive because the optical rotation caused by one enantiomer will be cancelled out by its non-superimposable mirror image. Such mixtures are denoted by prefix dl (±). For example, the racemic mixture of lactic acid will be written as dl (±)-lactic acid or simply (±)-lactic acid. The two enantiomers in a racemate can be separated by physical and chemical methods. The separation of enantiomers from a racemate is referred as resolution.

6.4 Configuration and the *R-S* Convention of Nomenclature

The three-dimensional arrangement of atoms or groups around a stereogenic carbon is described as configuration. Enantiomers will, thus, have opposite configuration. The nomenclature of different configurations is done by using the Cahn-Ingold-Prelog rule regarding the priority of atoms or groups around the stereocenter. Before describing this system, it would be appropriate for you to understand how to represent the structures of chiral molecules. This is explained below by taking lactic acid (**Figure 6.10**) as an example.

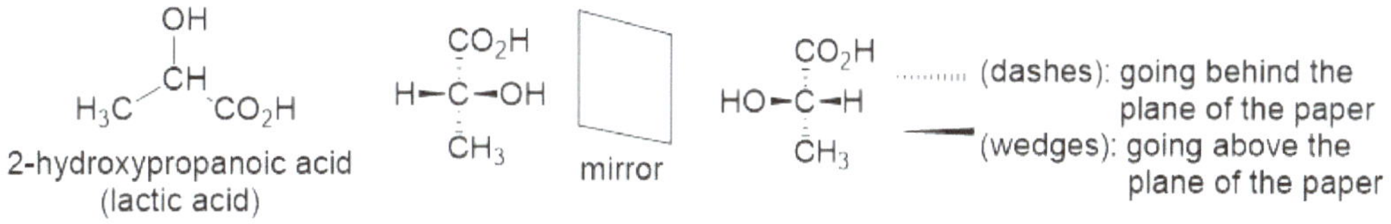

Fig. 6.10: *The dash – wedge formula of lactic acid.*

The dash – wedge formula to show the three-dimensional arrangement was simplified by Emil Fischer using a two-dimensional way that is called the Fischer Projection Formula (**Figure 6.11**). In this formula, carbon atom is not shown, and dashes and wedges are replaced with straight vertical and horizontal lines. The crossing point of the vertical and horizontal line represents the carbon atom. The vertical line leads the groups that project below the plane of the paper whereas the horizontal line projects the groups above the plane of the paper.

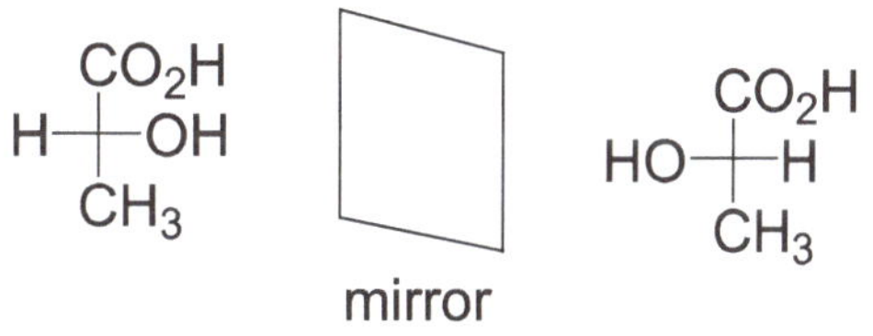

Fig. 6.11: *The Fischer projection formula of lactic acid.*

Another way of representation is perspective formula (**Figure 6.12**). In this formula, two of the bonds are shown in the plane of the paper, one bond as dashed line directing below the plane of the paper and the fourth bond as the wedged bond forwarding towards the viewer. The dashed and wedged bonds must be adjacent to each other and so the two-line bonds in the plane of the paper. For example, see below the two enantiomers of lactic acid.

Fig. 6.12: *Perspective formula of lactic acid.*

To describe the structure correctly, we need a system of nomenclature that indicates the arrangement of the atoms or groups around the stereocenter. Now a days *R,S* system of nomenclature is widely accepted to describe the configuration of enantiomers. In this system, four groups attached to a chiral carbon are prioritized as 1 (highest priority), 2, 3, and 4 (lowest priority) by the Cahn-Ingold-Prelog rule. The chiral carbon is then observed from the side opposite to the lowest priority group (4). If the remaining three groups 1, 2, and 3 form a clockwise pattern, the configuration is designated *R* (from the Latin *rectus* meaning "right"). If they are in anticlockwise fashion, the configuration is designated as *S* (from the Latin *sinister* meaning "left").

The four groups are prioritized using the following set of rules.

Rule 1

The atoms directly attached to the stereogenic center are ranked according to

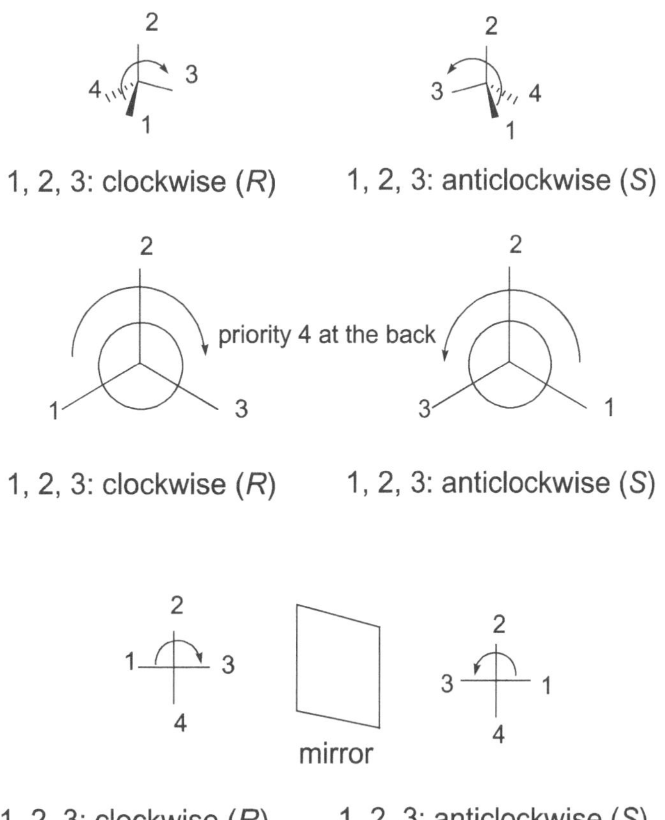

atomic number – the higher the atomic number, the higher the priority. If one of the four groups is H it is obviously the lowest priority (4), and the stereogenic carbon is viewed looking down the C-H bond from C to H. For example, if a chiral carbon is attached to a bromine atom, an oxygen atom, a hydrogen atom and a carbon atom, the priority of atoms/groups will be as follows.

Br > O > C > H

1 2 3 4

Rule 2

If two or more directly attached atoms are same, work outward from the stereogenic carbon until a decision is reached. In the molecule shown below, the priority of methyl and ethyl groups cannot be decided by comparing the 1st atom attached to stereogenic carbon because in both it is carbon atom that is directly attached to the stereogenic carbon. On working outward, we see that the ethyl group has higher priority than the methyl group because at the first point of difference between the methyl and ethyl groups there is a carbon atom in the ethyl group and a hydrogen atom in the methyl group.

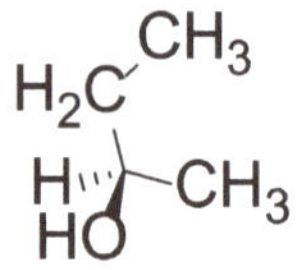

CH_3
H_2C
H — CH_3
HO

Rule 3

If a double or triple bond needs to be considered in determining the priority, the atoms involved are treated as being duplicated or triplicated, respectively. See the examples below.

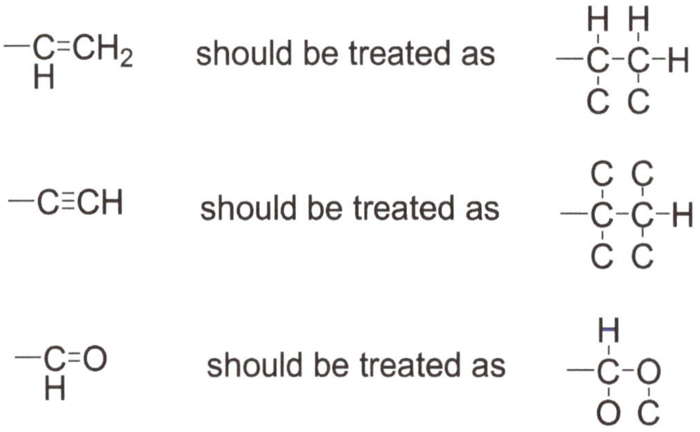

$-\underset{H}{C}=CH_2$	should be treated as	$-\underset{C}{\overset{H}{C}}-\underset{C}{\overset{H}{C}}-H$
$-C\equiv CH$	should be treated as	$-\underset{C}{\overset{C}{C}}-\underset{C}{\overset{C}{C}}-H$
$-\underset{H}{C}=O$	should be treated as	$-\underset{O}{\overset{H}{C}}-\underset{C}{O}$

Let's assume that a vinyl group and an isopropyl group are attached to the stereogenic carbon. The vinyl group will have priority over the isopropyl group because at the first point of difference there is a carbon atom in the vinyl group but a hydrogen atom in the isopropyl group.

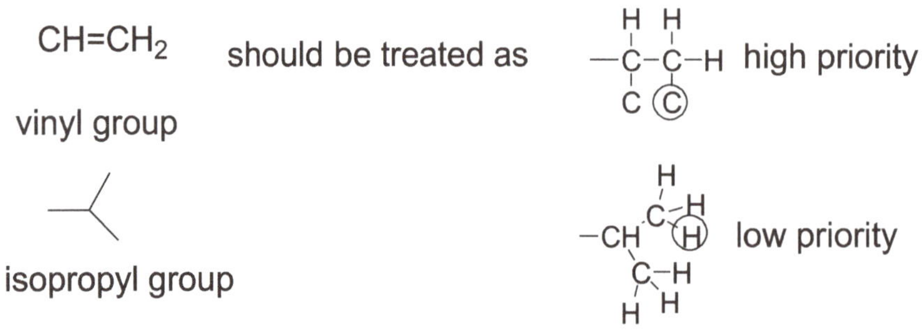

$CH=CH_2$ vinyl group	should be treated as	$-\underset{C}{\overset{H}{C}}-\underset{Ⓒ}{\overset{H}{C}}-H$	high priority
isopropyl group		$-CH(CH_3)(CH_3)$ with H, H, Ⓗ on one C and H, H, H on the other	low priority

Problem: Assign priority to the atoms or groups in the following sets.

a. -H, $-CH_2CH_3$, -Cl, $-CH_2NH_2$ b. –OH, -Br, $-CH_2CH_3$, $-CH=CH_2$

c. $-CH_3$, -COOH, -SH, $-CH_2Br$ d. $-CH_3$, $-NH_2$, $-NHCH_3$, -COOH

Solution: a. Cl, 2. CH_2NH_2, 3. CH_2CH_3, 4. H; b. 1. Br, 2. OH, 3. $CH=CH_2$, 4. CH_2CH_3; c. 1. SH, 2. CH_2Br, 3. COOH, 4. CH_3; d. 1. $NHCH_3$, 2. NH_2, 3. COOH, 4. CH_3.

Problem: Using the Cahn-Ingold-Prelog convention, assign *R* or *S* configuration to the following structures of 1-bromoethanol, 2-bromobutane and lactic acid.

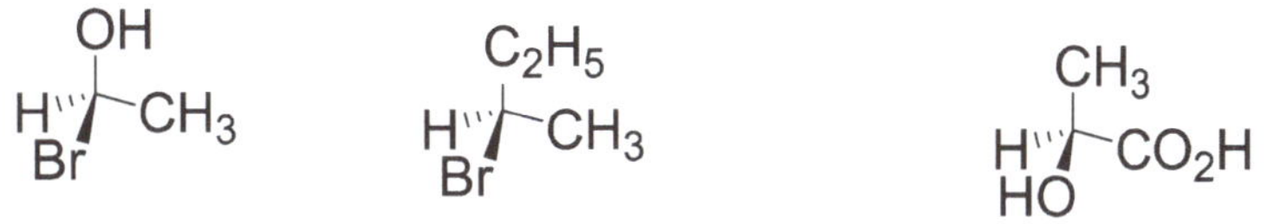

1-bromoethanol 2- bromobutane lactic acid

Solution: First, decide the priority of all the groups attached to the stereogenic carbon and then see from the side opposite to the least priority group (H in this case) if 1st, 2nd, and 3rd priority groups are in clockwise or anticlockwise pattern.

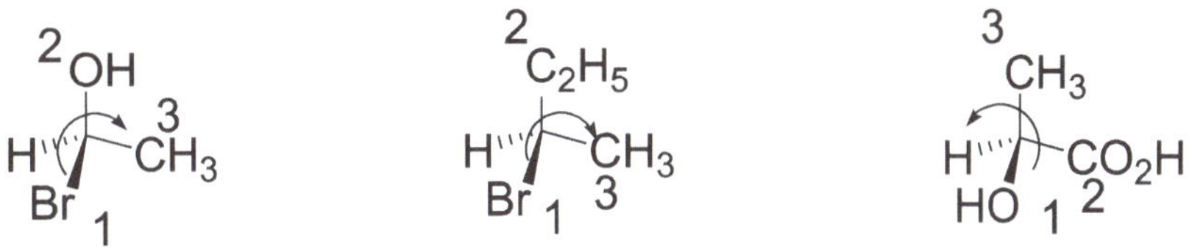

clockwise: (*R*) clockwise: (*R*) anticlockwise: (*S*)

(*R*)-1-Bromoethanol (*R*)-2-Bromobutane (*S*)-Lactic acid

Sometimes it may not be easy for students to view the structure in three dimensions on the peace of the paper. Students are therefore encouraged to use molecular model. One can decide the priority of groups on the piece of the paper and then fix those priority numbers on the appropriate bonds of the molecular model using different color balls. For example, see the structure below that is of lactic acid. This structure is obtained by exchange in positions of OH and H in (*S*)-lactic acid. Now you may not be able to view the clockwise pattern of the atoms or groups attached to the chiral carbon on the piece of the paper. You can only find a clockwise pattern when you are able to view the structure from the side opposite to the hydrogen atom (that is coming towards you in the structure). Keep in mind that if you are exchanging the position of two groups on a particular enantiomer you will get structure of its enantiomer. Two exchanges will bring the same structure.

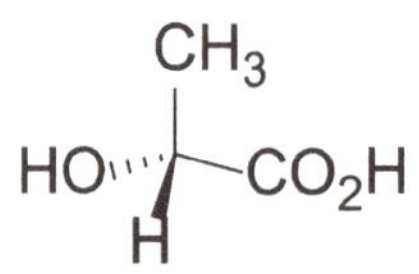

CH_3
HO····C—CO_2H
H

(R)-Lactic acid

SUMMARY

The compounds that rotate the plane of the plane-polarized light are called optically active compounds. The optical rotation by a compound can be measured with a polarimeter. There are many compounds with identical structure but differing only in three-dimensional arrangement of atoms or groups at one or more carbon atoms and rotate the plane of the plane-polarized light differently. Such compounds are called *optical isomers*. The origin of optical isomerism is in *chirality* of the molecule. The compounds due to different reasons might be *chiral* or *achiral*. All those compounds which contain one carbon atom that is attached to four different atoms or groups are chiral. These compounds exist as mirror images that are non-superimposable – mirror image isomers or *enantiomers*. All stereoisomers that are not mirror images are called *diastereomers*. A mixture of two enantiomers in equal amounts is known as *racemic* mixture. The compounds with a plane of symmetry – a plane that divides the molecule into two equal halves, are optically inactive and known as *meso*-compounds. The different configurations of the enantiomers are named by Cahn-Ingold-Prelog convention.

EXERCISES

1. Define the following terms.

   a. chiral molecule

   b. plane-polarized light

   c. enantiomers

   d. diastereomers

   e. racemic mixture

   f. resolution

2. What do you mean by optical rotation? A 1.50 g sample of cocaine, $[a]_D$ = - 16°, was dissolved in 10.00 mL of chloroform and placed in a sample tube with a path length of 5.00 cm. Calculate the observed rotation.

3. Which of the following compounds have stereogenic center(s)?

   a. 2,3-dimethylhexane

   b. 1,2-dibromopropane

   c. 2,2-dibromobutane

   d. 2-methylhexane

4. Draw structural formula for an optically active compound corresponding to each of the following molecular formula.

   a. C_4H_9Cl b. $C_4H_{10}O$ c. $C_5H_{13}N$ d. $C_5H_{11}Br$

5. Ephedrine is a drug used to treat asthma. Identify the chiral centers in ephedrine. How many stereoisomers are possible for ephedrine? Draw structure of all stereoisomers using dashes and wedges to clearly elucidate different configuration.

ephedrine

6. Assign configuration according to *R-S* convention to chiral centers in chloramphenicol which is a broad-spectrum antibiotic that is particularly used in treatment of typhoid fever.

chloramphenicol

7. The following compound has only one chiral carbon. However, it exists in four stereoisomeric forms. Draw structure of all four stereoisomers.

8. Assign *R* or *S* configuration to the following compounds.

(a) (b) (c) (d) (e)

9. Identify the following pairs of compounds as enantiomers or same compounds.

(a) Fischer projection: COOH (top), H (left), NH_2 (right), CH_3 (bottom) and Fischer projection: NH_2 (top), H (left), COOH (right), CH_3 (bottom)

(b) Fischer projection: SH (top), H (left), CH_3 (right), C_2H_5 (bottom) and Fischer projection: C_2H_5 (top), H (left), CH_3 (right), SH (bottom)

10. What would you expect about the optical activity of the following compound? Explain.

HOOC–C(H)(Cl)–C(H)(Cl)–COOH (Fischer-type projection: H on top of each carbon, Cl at bottom of each carbon)

7 ORGANIC HALOGEN COMPOUNDS: NUCLEOPHILIC SUBSTITUTION AND ELIMINATION REACTIONS

Organic halogen compounds, especially those containing chlorine and bromine are widespread in nature. Many species that live in sea such as sponges and mollusks are sources of organic halogen compounds. Chloromethane is released from forest fires, volcanoes, and from ocean kelp.

Organic halogen compounds, especially those containing chlorine and bromine are also versatile reagents in organic synthesis. The halogen atoms can be replaced by various nucleophiles to form other classes of organic compounds. The dehydrohalogenation of alkyl halides leads to the formation of alkenes. Many organic halogen compounds containing two or more halogen atoms are used as solvents for chemical reactions, as insecticides and as herbicides, cleaning fluids, refrigerants, and polymers (**Figure 7.1**).

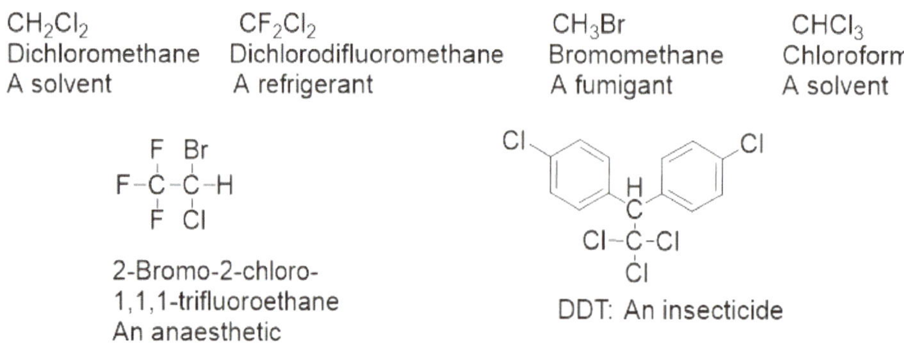

Fig. 7.1: *Some useful organic halogen compounds.*

7.1 Preparation, Nomenclature and Classification

The formation of some organic halogen compounds has been described previously by free-radical substitution reactions of alkanes, electrophilic aromatic substitution reactions of benzene, and by addition of halogens or hydrogen halides on carbon-carbon double bond of alkenes and carbon-carbon triple bond of alkynes. Organic halogen compounds can also be prepared by substitution of –OH group of alcohols using hydrogen halides or $SOCl_2$ or PCl_3 (see Chapter 8). Although organic halogen compounds can be aliphatic or aromatic we will focus on nucleophilic substitution and elimination reactions of alkyl halogen compounds (alkyl halides or haloalkanes, e.g., bromo for bromine, chloro for chlorine) and their mechanisms in this chapter. At least some reactions of aromatic halogen compounds you have seen during discussion on electrophilic aromatic substitution reactions of benzene derivatives.

For naming the haloalkanes, select the longest possible chain as the parent chain. The parent chain is numbered from the side that gives minimum number to the first substituent, be it halo or alkyl group (**Figure 7.2**). If a multiple bond is present, it must be taken in the parent chain. Clearly, a multiple bond has priority over halogens. Write the name using other rules as described for open-chain hydrocarbons.

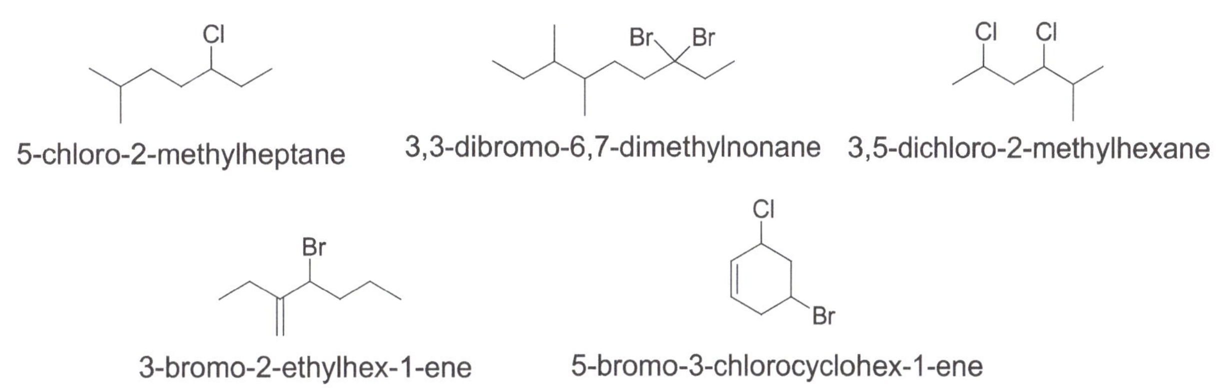

Fig. 7.2: *Naming alkyl halogen compounds (alkyl halides).*

Aliphatic (open-chain or cyclic) organic halogen compounds can be classified as primary (1°), secondary (2°) and tertiary (3°) depending on the number of carbon or hydrogen atoms attached to the carbon bearing halogen atom (**Figure 7.3**). If the carbon bearing halogen atom is attached to one carbon atom and two hydrogen atoms the organic halogen compound is called primary or 1° organic halogen compound. If the carbon bearing halogen atom is attached to two carbon atoms and one hydrogen atom the organic halogen compound is called secondary or 2° organic halogen compound. The tertiary or 3° aliphatic halogen compounds are those in which the carbon bearing the halogen atom is attached to three carbon atoms and hence no hydrogen atom attached to it.

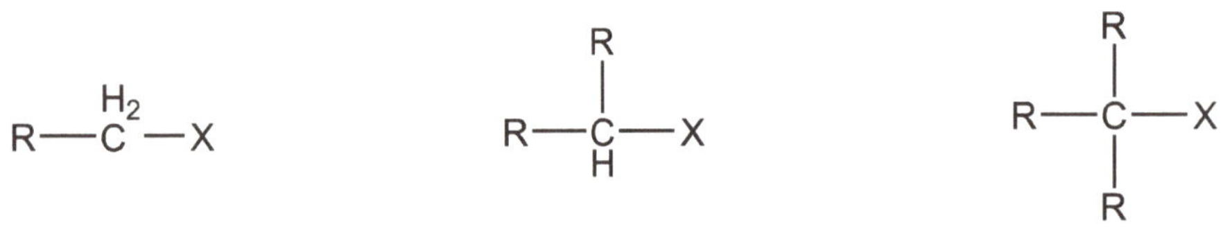

$R-CH_2-X$ $R-CH(R)-X$ $R-C(R)_2-X$

R = Aliphatic group
X = Halogen atom

Pr. aliphatic halogens   Sec. aliphatic halogens   Tert. aliphatic halogens

Fig.: 7.3: *Primary, secondary, and tertiary organic halogen compounds.*

The reactivity of organic halogen compounds in nucleophilic substitution and elimination reactions depends on whether they are primary, secondary, or tertiary.

Problem: Identify the following organic halides as primary, secondary, or tertiary.

Br (a.) H_3C Br (b.)   Br (c.)

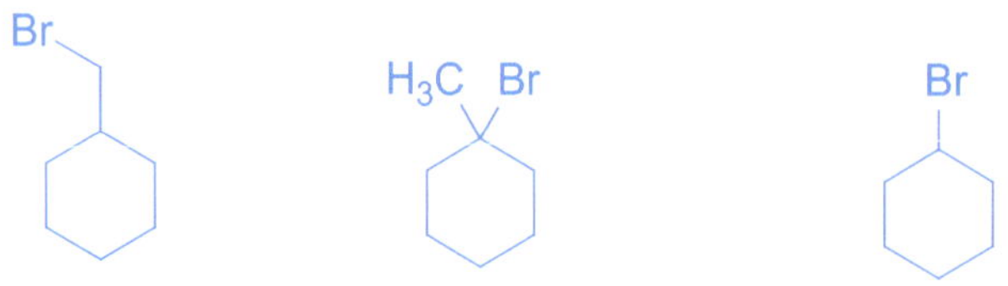

a.   b.   c.

Solution: a. primary, b. tertiary, c. secondary

7.2 Nucleophilic Substitution Reactions

Nucleophilic substitution reactions are one of the most important and widely investigated reactions of alkyl halides. The halogen atom is substituted, or you can say replaced as halide ion by a variety of nucleophilic reagents to form useful products. Some examples are shown below. You must keep in mind that the nucleophile might be negatively charged ion or neutral molecule.

$$H_3C-CH_2-Br + NaOH \xrightarrow{H_2O} H_3C-CH_2-OH + NaBr$$

OR

$$H_3C-CH_2-Br + \overset{\ominus}{O}H \xrightarrow{H_2O} H_3C-CH_2-OH + Br^{\ominus}$$

The nucleophilic substitution reactions can be generalized as shown below.

$$R{-}X + \ddot{N}u^{\ominus} \longrightarrow R{-}Nu + X^{\ominus}$$

ionic nucleophile

$$R{-}X + \ddot{N}u \longrightarrow R{-}Nu^{\oplus} + X^{\ominus}$$

neutral nucleophile

In the case of neutral nucleophiles, the cationic product formed is stabilized by removal of a proton. Some common nucleophiles and organic products obtained from their reaction with alkyl halides (R-X) are listed below (**Table 7.1**).

Oxygen Nucleophiles	Product
OH^-	R-OH (alcohol)
RO^-	R-O-R (ether)
H_2O	$R\text{-}OH_2^+$ $\xrightarrow{-H^+}$ R-OH (alcohol)
R-OH	R_2OH^+ $\xrightarrow{-H^+}$ R-O-R (ether)
$R'\text{-}COO^-$	R-O-COR' (ester)

Nitrogen Nucleophiles | Product

$$R{-}X \xrightarrow[-X^{\ominus}]{NH_3} R{-}NH_3^+ \xrightarrow{-H^+} R{-}NH_2 \text{ (1° amine)}$$

$$R{-}X \xrightarrow[-X^{\ominus}]{R\text{-}NH_2} R_2NH_2^+ \xrightarrow{-H^+} R_2NH \text{ (2° amine)}$$

$$R{-}X \xrightarrow[-X^{\ominus}]{R_2NH_2} R_3NH^+ \xrightarrow{-H^+} R_3N \text{ (3° amine)}$$

$$R{-}X \xrightarrow[-X^{\ominus}]{R_3NH_2} R_4N^+ \text{ (quaternary ammonium ion)}$$

Sulfur Nucleophiles	Product
SH^-	R-SH (thiol)
RS^-	R-S-R (thioether)

Carbon Nucleophiles	Product
$^{\ominus}C\equiv N$ cyanide	$R-C\equiv N$
$^{\ominus}C\equiv CH$ acetylide	$R-C\equiv H$

Table 7.1: *Common nucleophiles and their reaction product with alkyl halides.*

Problem: How can you prepare i) cyclohexanol from bromocyclohexane, ii) 1-butyne from bromoethane using nucleophilic substitution reaction.

Solution: i) The reaction of bromocyclohexane with aqueous sodium hydroxide gives cyclohexanol, ii) the reaction of bromoethane with sodium acetylide will give 1-butyne.

i) Bromocyclohexane $\xrightarrow{\text{NaOH (aq.)}}$ cyclohexanol (OH)

ii) $HC\equiv CH \xrightarrow[\text{liquid } NH_3]{NaNH_2} HC\equiv C^{\ominus}\,Na^+$ (sodium acetylide) $\xrightarrow{CH_3CH_2Br} HC\equiv C-CH_2-CH_3$

a. Mechanisms of Nucleophilic Substitutions

There are two main mechanisms of nucleophilic substitution. These mechanisms are described by symbols S_N2 and S_N1.

The S_N2 mechanism

The S_N2 Mechanism involves only one step process. The letters S and N in the name of mechanism denote substitution and nucleophilic, respectively, whereas the digit "2" stands for bimolecular nature of the reaction. The key step of the reaction, also called the rate determining step, involves two molecules and is bimolecular.

The reaction takes place by attack of the nucleophile on substrate from the back side of the leaving group (halide). The nucleophile gets partially attached to the carbon atom bearing the leaving group which simultaneously gets partially detached. The species with partially bonded nucleophile and halogen is called transition state. As the leaving group departs with its electron pair, the nucleophile gets completely attached to the carbon using its electron pair.

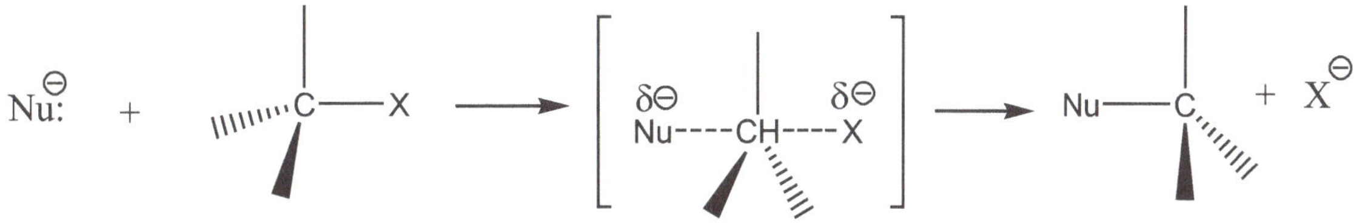

Transition State

Evidences in support of the S_N2 mechanism:

1. The rate of reaction depends on concentration of both the substrate and nucleophile. It supports the bimolecular nature of the reaction. The reaction of ethyl bromide with hydroxide nucleophile is an example of reaction taking place by S_N2 mechanism. If the concentration of either alkyl halide or nucleophile is changed, the rate of the reaction changes in proportion to that change.

   Rate = k $[C_2H_5Br][HO^-]$

2. S_N2 displacement occurs with inversion of configuration of the product. Thus, if halide is having (*R*)-configuration the product will have the (*S*)-configuration and vice-versa. The inversion of configuration requires attack of nucleophile from the side opposite to leaving group.

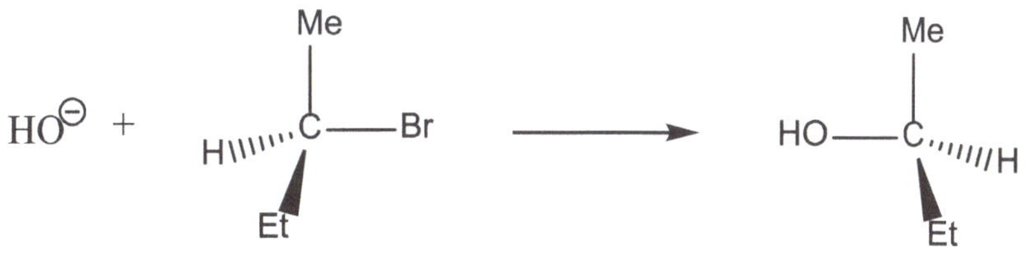

3. The reaction is fastest with primary halides and slowest with tertiary halides. Secondary alkyl halides react at an intermediate rate. It is because the rear side of the halogen bearing carbon is the most sterically hindered in case of 3° halides and least hindered in 1° halides. Steric hindrance is nothing but the presence of more groups or any bulky group in route of nucleophile to carbon bearing the halogen atom that hinders the approach of the nucleophile to this carbon atom.

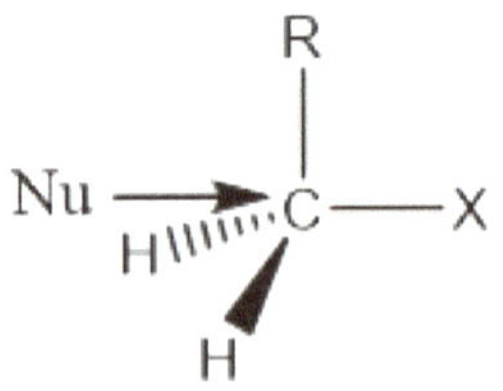

(Primary halide, rear side not crowded: S_N2 fast)

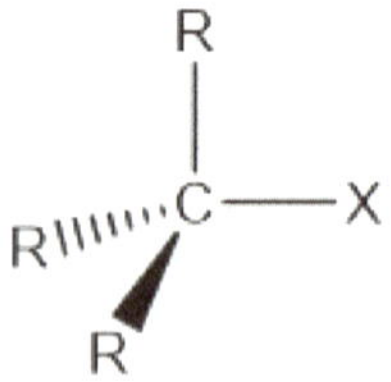

(Tertiary halide, rear side crowded: S_N2 very slow or impossible)

Problem: The reaction of $CH_3CH_2CH_2Br$ with sodium cyanide in acetone to give $CH_3CH_2CH_2CN$ proceeds by S_N2 mechanism. What will be the effect on rate of reaction if the concentration of both the substrates is doubled?

Solution: The reaction rate will increase by a factor of four.

The S_N1 mechanism

The S_N1 Mechanism involves a two-step process. The key step or the rate-determining step of the reaction involves only alkyl halide and hence the mechanism is unimolecular (1 in S_N1 stands for unimolecular). In this step, which is the first step of the reaction, the bond between the carbon and halogen breaks as the substrate ionizes. The electron pair of the C-X bond goes with the leaving group (halide ion), and a carbocation is formed. In the second step of the reaction, the carbocation combines with the nucleophile to yield the final product (if the nucleophile is ionic).

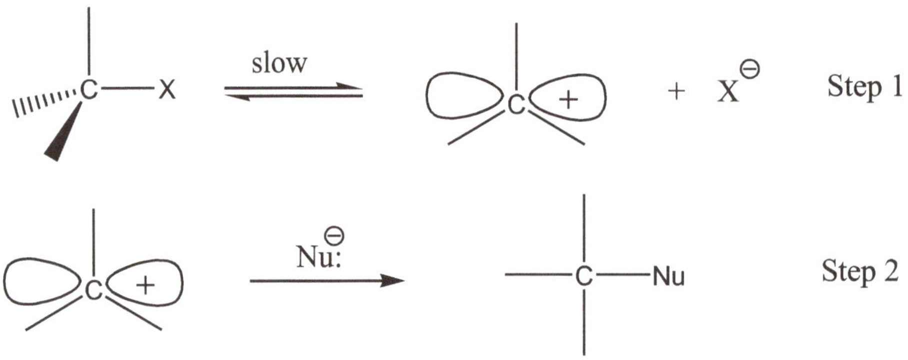

Step 1

Step 2

If the nucleophile is a neutral molecule such as water or an alcohol, its reaction with the carbocation would lead to an ionic species which will lose a proton from the nucleophilic center in the third step to give the final product.

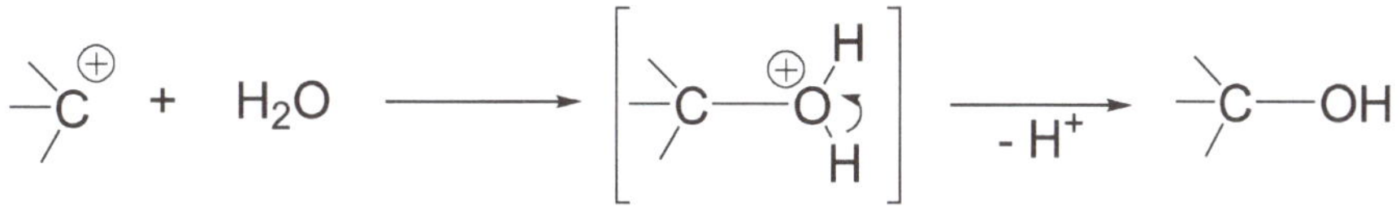

Evidences in support of the S_N1 mechanism

1. The rate of the reaction is independent of concentration of the nucleophile. For example, the rate of reaction in the reaction of *tert*-butyl bromide with water was observed to be proportional of the molar concentration of the *tert*-butyl bromide. It supports the fact that the 1st step that is ionization and does not involve nucleophile is the key step of the reaction.

   Rate = $k[(CH_3)_3CBr]$

2. If the carbon bearing the halogen is chiral and the substrate is optically pure, the reaction by S_N1 mechanism leads to the formation of racemic mixture of product.

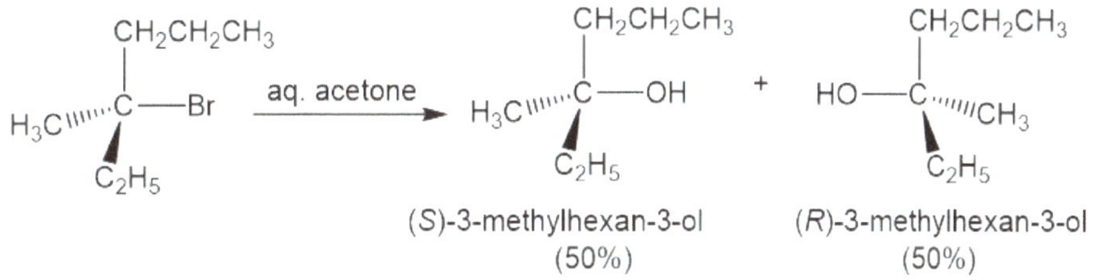

(*S*)-3-methylhexan-3-ol (50%) (*R*)-3-methylhexan-3-ol (50%)

What is the explanation for the formation of racemic mixture? In carbocations, only three groups are attached to the positively charged carbon atom. Therefore, it is sp^2-hybridized and planar. The nucleophile can approach the cationic carbon from either face of it to give a 50/50 mixture of the two enantiomers (a racemic mixture).

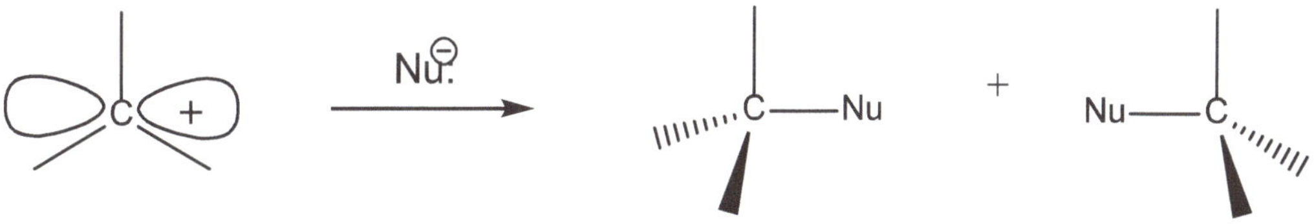

3. The reaction is fastest with tertiary halides and slowest with primary halides. This observation supports the formation of carbocation in the key step. The reaction rate is faster in case of tertiary halides because the carbocation formed

in this case would be a tertiary carbocation (3° carbocation). Such carbocations are more stable in comparison to secondary (2°) or primary (1°) carbocations.

4. Allylic halides also undergo nucleophilic substitution by S_N1 mechanism. In this case, the carbocation formed is stabilized by resonance.

$$\left[H_3C-\overset{\oplus}{C}H-CH=CH_2 \longleftrightarrow H_3C-CH=CH-\overset{\oplus}{C}H_2 \right]$$

Problem: The reaction of *tert*-butyl bromide, $(CH_3)_3C\text{-}Br$ with aqueous sodium hydroxide to give *tert*-butyl alcohol $(CH_3)_3C\text{-}OH$ proceeds by S_N1 mechanism. What will be the effect on rate of reaction if the concentration of i) the base is doubled, ii) the concentration of *tert*-butyl bromide is reduced to half?

Solution: i) The reaction rate will not change because there is no nucleophile in rate-determining step. ii) The reaction rate will also reduce to half because it is directly proportional to the molar concentration of the bromide,

b. Factors affecting S_N1 and S_N2 mechanisms

1. The structure of halides

The primary halides react by S_N2 mechanism and tertiary halides react by S_N1 mechanism. The secondary halides react by both mechanisms.

2. The effect of solvents:

The polar protic solvents (R-OH, H_2O) favor S_N1 mechanism. Such solvents solvate the ions and so enhance S_N1 reactivity. On the other hand, solvation of nucleophiles ties up their electron pairs. Therefore, the reaction by S_N2 mechanism, where the rate depends upon nucleophile effectiveness, is retarded by polar protic solvents. Polar aprotic solvents such as acetone, DMSO, DMF, accelerate the rate of reactions that take place by S_N2 mechanism.

3. The effect of nucleophiles:

The rate of a reaction occurring by S_N2 mechanism depends on the nucleophile. A stronger nucleophile favors S_N2 mechanism. The following guidelines are followed in determining the strength of the nucleophiles:

i. Nucleophiles with negative charge are stronger or better electron suppliers than the corresponding neutral nucleophiles. Some examples are shown below.

$HO^- > H_2O$, $RS- > R-SH$, $RO^- > R-OH$

ii. Elements low in the periodic table tend to be more nucleophilic than elements above them in same column.

HS- > HO-, $I^- > Br^- > Cl^- > F^-$

iii. In a period, the more electronegative the element, the weaker is the nucleophile. For example,

$R_3C^- > R_2N^- > RO^- > F^-$

4. The Leaving groups

Another factor that can affect the nucleophilic substitution reactions is the nature of the leaving group. The best leaving groups are those that give the most stable anions. Since the stability of anions is inversely related to basicity, the weakest base (anions of strong acids) will be the most stable anion and hence the best leaving group. The relative reactivity of the halides is as follows.

F^-	< Cl^- <	Br^-	< I^-	<	TsO^-
1	200	10,000	30,000	60000	

Problem: Which of the following bromides will react faster with ethanol (S_N1)?

a. $CH_3CH_2C(CH_3)_2Br$ or $CH_3CH_2CH(CH_3)Br$

b. $CH_3CH_2CH_2Br$ or $H_2C{=}CHCH_2Br$

Solution:

a. Of the two bromides the former one is a tertiary bromide while the latter is a secondary bromide. Hence, the former will react faster in S_N1.

b. Of the two bromides the former one is a primary bromide while the latter is an allylic bromide. In case of allylic bromide, the carbocation will be stabilized by resonance. Hence, it will react faster in SN1 that proceeds via carbocation.

7.3 Elimination Reactions

Alkyl halides undergo elimination reactions, a very useful method for preparation of alkenes. If the carbon atom adjacent to the carbon atom bearing halogen in an alkyl halide is attached to a hydrogen atom, competing reactions of alkyl halide with a nucleophile are possible. For example, in the scheme shown below, the first reaction is nucleophilic substitution in which the nucleophile is replacing the halogen atom as halide. In the second reaction, the nucleophile is working as a base by abstracting the proton. This reaction of alkyl halide with nucleophile forming an alkene is an elimination reaction. The symbol E denotes an elimination reaction.

H–C–C–X + Nu:

Substitution → H–C–C–Nu + $X^{\ominus}$

Elimination → C=C + NuH + $X^{\ominus}$

Further, in unsymmetrical alkyl halogen compounds like 2-bromobutane that contains hydrogen atoms on both carbon atoms adjacent to the carbon bearing halogen, there are two elimination products possible. In such cases, the more stable alkene, that is more substituted at C=C, is the major product according to Zaitsev's rule. For example, 2-butene is the major product in the reaction of 2-bromobutane with sodium ethoxide because both doubly bonded carbon atoms are attached to one hydrogen atom and one methyl group. In 1-butene, the terminal *sp2* carbon atom has two hydrogen atoms meaning it is not substituted.

$$\text{2-bromobutane (Br)} \xrightarrow[CH_3CH_2OH]{CH_3CH_2O^-Na^+} \underset{\text{2-butene (81\%)}}{H_3CHC{=}CHCH_3} + \underset{\text{1-butene (19\%)}}{H_3CH_2CHC{=}CH_2}$$

Problem: Write all possible elimination products from the following reaction.

$$\text{(Br)} \xrightarrow[CH_3CH_2OH]{CH_3CH_2O^-Na^+}$$

Solution:

$$\text{(Br)} \xrightarrow[CH_3CH_2OH]{CH_3CH_2O^-Na^+} \text{(alkene)} + \text{(alkene)} + \text{(alkene)}$$

Mechanism of Elimination Reactions

Like nucleophilic substitution reactions, there are two main mechanisms for elimination reaction – E2 mechanism and E1 mechanism.

The E2 mechanism

The E2 mechanism, like S_N2 mechanism, is a one – step process. The nucleophile, which acts as a base, removes the proton (hydrogen without its bonding electron pair) from the carbon atom adjacent to one bearing the leaving group. At the same time, the leaving group departs and a carbon – carbon double bond (C=C) is formed.

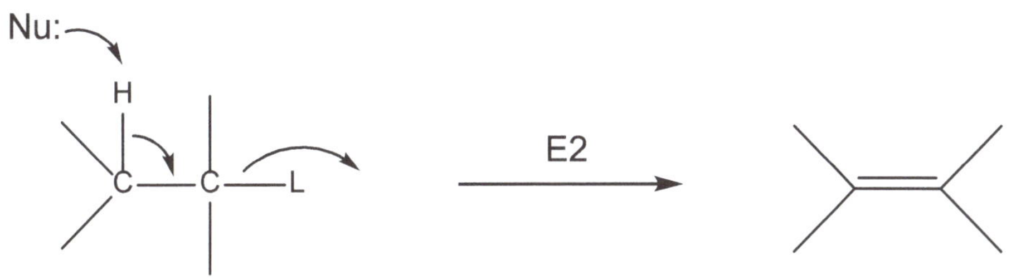

Nu: H C C L E2

The E1 mechanism

The E1 mechanism is a two – step process and has the same first step as the S_N1 mechanism – ionization of substrate to form carbocation. The carbocation then loses the proton to form the alkene.

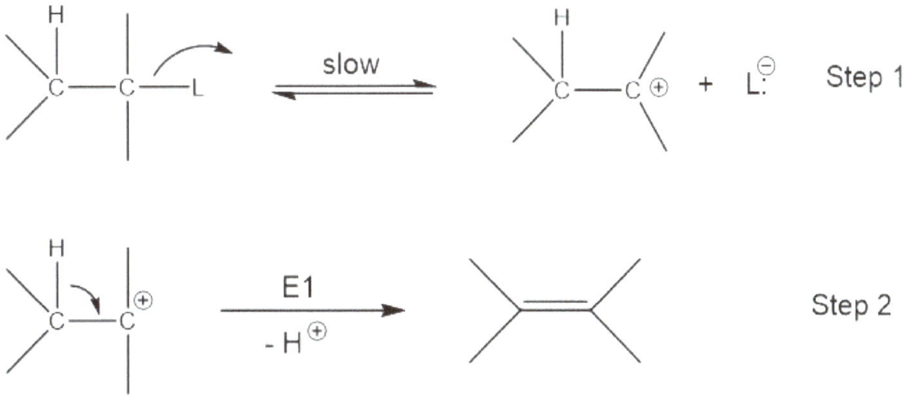

H C C L slow H C C ⊕ + L:⊖ Step 1

H C C ⊕ E1 - H⊕ Step 2

7.4 Nucleophilic Substitution versus Elimination

Primary Halides

i. Primary halides undergo nucleophilic substitution and elimination reactions only by S_N2 and E2 mechanisms, respectively. It is because the S_N1 and E1 mechanisms go through carbocations and ionization of primary halides to carbocations do not occur.

ii. Primary halides react with most nucleophiles giving S_N2 product. E2 product is favored only with very bulky, strongly basic nucleophiles. For example, 1-bromobutane undergoes substitution with sodium ethoxide and elimination with potassium *tert*-butoxide.

EtONa / EtOH: (S_N2, 90%) + (E2, 10%)

t-BuOK / *t*-BuOH: (S_N2, 15%) + (E2, 85%)

Tertiary Halides

i. In case of tertiary halides, substitution reactions occur only by S_N1 and not by S_N2 mechanism, but elimination reactions occur by both E1 and E2 mechanisms.

ii. Weak nucleophiles and polar solvents favor S_N1 product over E1 product.

tert.-butyl bromide $\xrightarrow{H_2O}$ $(CH_3)_3C^{\oplus}$ + $Br^{\ominus}$

H_2O, S_N1: $(CH_3)_3C-OH$ (80%)

$-H^{\oplus}$, E1: $(CH_3)_2C=CH_2$ (20%)

iii. Strong nucleophiles and less polar solvents favor elimination by E2 mechanism.

$$(CH_3)_3C\text{-}Br \xrightarrow[E2]{HO^{\ominus}} (CH_3)_2C{=}CH_2$$

Secondary Halides

i. The reactions of secondary halides take place by all four mechanisms S_N2, S_N1, E2 and E1.

ii. Stronger nucleophiles that are not stronger bases (e.g., $CH_3CH_2S^-Na^+$) favor substitution by S_N2 mechanism whereas weaker nucleophiles such as polar solvents (e.g., C_2H_5OH) favor substitution by S_N1 mechanism.

$$(CH_3)_2CH\text{-}Br + CH_3CH_2S^{\ominus}Na^{\oplus} \xrightarrow{S_N2} (CH_3)_2CH\text{-}SCH_2CH_3$$

$$(CH_3)_2CH\text{-}Br + CH_3CH_2OH \xrightarrow[\text{(Major)}]{S_N1} (CH_3)_2CH\text{-}OCH_2CH_3$$

iii. A strong base (e.g., $CH_3CH_2O^-Na^+$) favors elimination reaction by E2 mechanism.

$$(CH_3)_2CH\text{-}Br + CH_3CH_2O^-Na^+ \xrightarrow[\text{(Major)}]{E2} H_3C\text{-}CH{=}CH_2$$

7.5 Formation of Grignard Reagents

Organic halogen compounds, whether aliphatic or aromatic, react with metals to form compounds known as organometallic compounds. Organometallic compounds have wide ranging applications in organic synthesis. The reaction of organic halogen compounds with magnesium forms organic magnesium halides known as the Grignard reagents after the name of its discoverer Victor Grignard. For example, ethyl bromide reacts with magnesium in anhydrous ether to form ethyl magnesium bromide. The Grignard reagents are highly prone to protonation and oxidation. Therefore, its preparation must be carried out in an anhydrous solvent under nitrogen or argon atmosphere. Usually, it is prepared in anhydrous diethyl ether or tetrahydrofuran (THF, a cyclic ether) and reacted with other reagents in the same flask to prepare different types of organic compounds. The Grignard reagents are, thus, valuable reagents for one-pot synthesis of various organic compounds.

$$C_2H_5{-}Br \quad + \quad Mg \xrightarrow{\text{ether}} C_2H_5MgBr$$

Since magnesium is a strongly electropositive element, the Grignard reagents behave like magnesium salt ($R^{-}\,{}^{+}MgBr$). It is, therefore, an important reagent for generating carbon nucleophile and forming carbon-carbon single bond. It, however, does not react with organic halides. On reaction with hydrogen halides, water or alcohols, the Grignard reagents easily get converted to hydrocarbons by accepting a proton. Thus, the reaction of Grignard reagents with water is an important method for conversion of organic halogen compounds to hydrocarbons.

$$R{-}Br + Mg \xrightarrow[\text{ether}]{\text{dry}} R{-}Mg{-}Br \xrightarrow{H{-}OH} R{-}H + Mg(OH)Br$$

Another important reaction of the Grignard reagents is their nucleophillic addition to C=O bond that will be discussed in the chapter on carbonyl compounds.

SUMMARY

1. Reaction Summary

Alkyl halides react with a variety of nucleophiles forming nucleophilic substitution products such as alcohols, ethers, alkynes, alkyl halides, amines, and nitriles, etc. Nucleophiles can be either negatively charged or neutral species.

$$R\text{-}X + Nu:^{\ominus} \longrightarrow R\text{-}Nu + X^{\ominus}$$

$$R\text{-}X + Nu: \longrightarrow R\text{-}Nu^{\oplus} + X^{\ominus}$$

In addition to undergoing nucleophilic substitution reactions alkyl halides undergo elimination reactions to give alkenes. The major product of elimination reaction is more stable alkene (more branched).

$$H\text{-}C\text{-}C\text{-}X + B:^{\ominus} \longrightarrow C{=}C + BH + X^{\ominus}$$

The reaction of both aliphatic and aromatic halogen compounds with magnesium in anhydrous ether leads to the formation of the Grignard reagent which is a valuable reagent in organic synthesis.

2. Mechanism Summary

The nucleophilic substitution reactions occur by S_N2 or S_N1 mechanism. The S_N2 mechanism is a one-step mechanism involving a transition state while the S_N2 mechanism is a two-step mechanism involving a carbocation. The S_N2 reaction occurs with inversion of configuration while the S_N1 gives a racemic mixture. Usually, the primary halides undergo nucleophilic substitution by S_N2 mechanism, and tertiary halides by S_N1 mechanism. The secondary halides can react by both mechanisms depending on solvent and nucleophile used. Elimination reactions also occur by two general mechanisms E2 and E1.

S_N1 Mechanism

slow
+ $X^{\ominus}$
Step 1

$Nu^{\ominus}$:
Step 2

S_N2 Mechanism

$Nu^{\ominus}$ + C—X → [δ⊖ Nu---HC----X δ⊖] → Nu—C + $X^{\ominus}$

Transition State

E2 Mechanism

Nu:
H
C—C—L
E2

E1 Mechanism

H
C—C—L
slow
C—C⊕ + $L^{\ominus}$:
Step 1

E1
- $H^{\oplus}$
Step 2

EXERCISES

1. Write the IUPAC name of the following organic halogen compounds.

Br Cl Cl

Br Br Cl

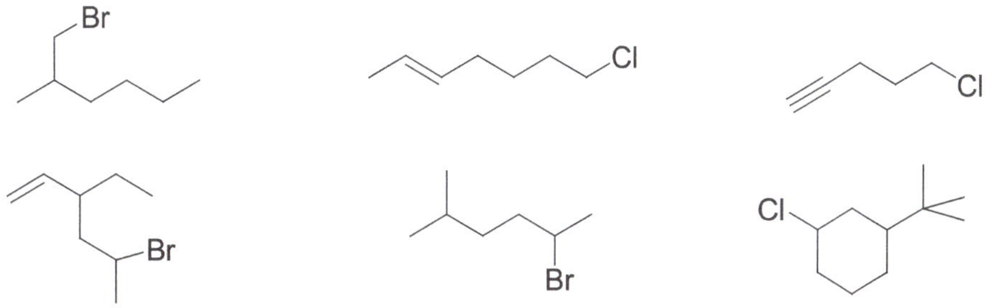

2. Draw the structure of (a) a primary alkyl iodide, C_3H_7I, (b) a secondary alkyl chloride, C_4H_9Cl, (c) a tertiary alkyl bromide, $C_6H_{13}Cl$.

3. Write structural formula and name of all possible alkyl halides having molecular formula C_4H_9Br.

4. Write the name, structure, and uses of three organic halogen compounds.

5. Complete the following chemical equations by providing the missing reactant or substitution product.

a.  Br + ? ⟶ OH

b.  Cl + C_2H_5ONa ⟶ ?

c. I + $HC{\equiv}C^{-}Na^{+}$ ⟶ ?

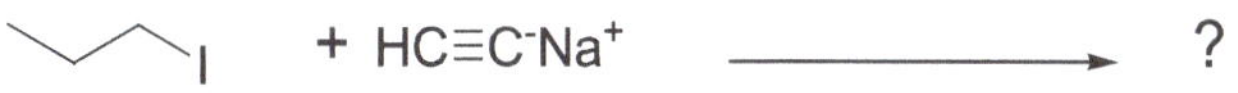

d.  $-CH_2Br$ + NaCN ⟶ ?

e.  Cl CH_3 + ? ⟶ SH CH_3

f.  Br + ? ⟶ NH_2

6. Show reaction equations to prepare following compounds by nucleophilic substitution of alkyl halides.

a. $CH_3CH(CH_3)CH(CH_3)NH_2$ b. $CH_3CH_2CH_2OCH_2CH_3$

c. 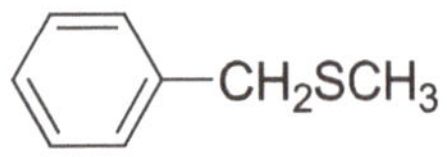 (benzene ring)$-CH_2SCH_3$ d. 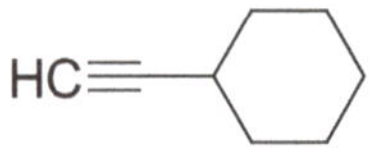 $HC\equiv$(cyclohexyl) e. 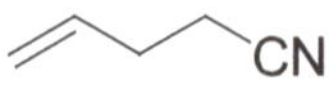 (but-3-enyl)CN

7. Draw the Fischer projection formula(s) for the product(s) of the following S_N1 reaction.

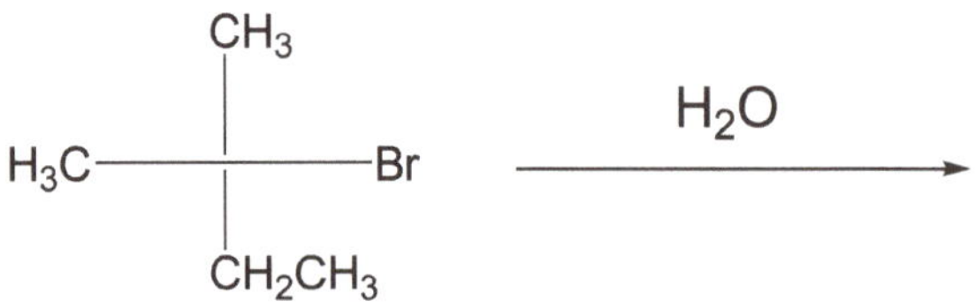

Fischer projection: top CH_3, left H_3C, right Br, bottom CH_2CH_3 $\xrightarrow{H_2O}$

8. (*R*)-2-Bromobutane reacts with sodium methoxide in methanol by S_N2 reaction. Write the chemical equation for this reaction.

9. Clearly show the mechanisms of S_N1 and S_N2 reactions in exercises 7 and 8.

10. Describe the effect of the following variables on both S_N1 and S_N2 reactions.
a. substrate structure, b. leaving group, c. solvents.

11. Arrange each series of compounds in their decreasing S_N2 reactivity.

a. CH_3CH_2Cl $(CH_3)_3Cl$ $CH_3CH_2CH(CH_3)Cl$

b. 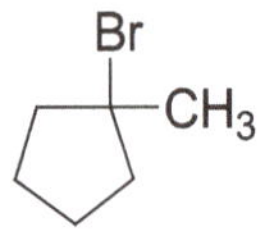 (1-bromo-1-methylcyclopentane: Br, $-CH_3$) (bromomethylcyclopentane: $-$Br) (bromocyclopentane: Br)

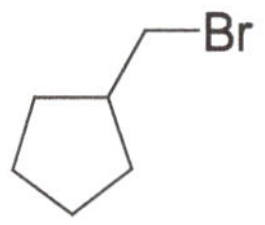

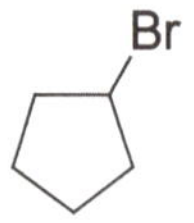

12. What effect would you expect on the S_N1 reaction of *tert*-butyl bromide with methanol (CH_3OH) forming *tert*-butyl methyl ether, $(CH_3)_3COCH_3$ if following changes are affected in the reaction?

a. The concentration of methanol is doubled.

b. The concentration of *tert*-butyl bromide is reduced to half.

c. The concentration of *tert*-butyl bromide is tripled and that of methanol is reduced to its one third.

d. The concentration of both *tert*-butyl bromide and methanol is doubled.

13. What effect would you expect on the S_N2 reaction of isopropyl bromide with sodium cyanide in acetone if following changes are affected in the reaction?

a. The concentration of sodium cyanide is tripled.

b. The concentration of *tert*-butyl bromide is doubled.

c. The concentrations of both *tert*-butyl bromide and sodium cyanide are doubled.

d. The concentration of *tert*-butyl bromide is doubled and that of sodium cyanide is reduced to half.

14. Explain the formation of different products in the following two reactions by considering their reaction mechanisms.

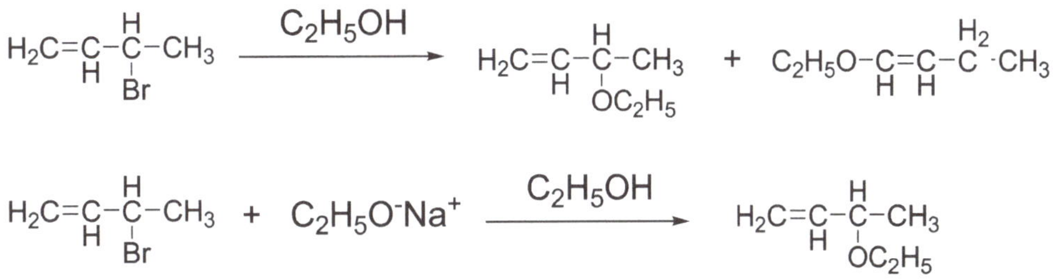

$$H_2C{=}CH{-}CH(Br){-}CH_3 \xrightarrow{C_2H_5OH} H_2C{=}CH{-}CH(OC_2H_5){-}CH_3 + C_2H_5O{-}CH{=}CH{-}CH_2{-}CH_3$$

$$H_2C{=}CH{-}CH(Br){-}CH_3 + C_2H_5O^-Na^+ \xrightarrow{C_2H_5OH} H_2C{=}CH{-}CH(OC_2H_5){-}CH_3$$

15. Select the stronger nucleophile in each pair.

a. NH_3 and $^-NH_2$, b. H_2S or ^-SH, c. CH_3COO^- or $CH_3CH_2O^-$

16. C_2H_5OH or C_2H_5O-, f. H_2O and HO^-, g. HO^- and ^-SH

17. Give the structure of all possible elimination products from following alkyl halides.

a. 2-bromo-2-methylpentane, b. 2-bromopentane, c. 3-bromo-3-methylhexane

18. Explain why 1-bromo-2,2-dimethylbutane has difficulty undergoing either S_N1 or S_N2. Can it undergo elimination reaction?

19. Identify the following reactions as either S_N1, S_N2, E1 or E2.

$$C_6H_5\text{-}CH(Br)\text{-}CH_3 \xrightarrow{\text{NaOH}} C_6H_5\text{-}CH{=}CH_2$$

$$C_6H_5\text{-}CH(Br)\text{-}CH_3 \xrightarrow{C_2H_5OH,\ \text{heat}} C_6H_5\text{-}CH(OC_2H_5)\text{-}CH_3$$

20. Which of the following two alkyl halide can undergo E1 more rapidly? Explain.

(cyclopentenyl)–Br or (cyclopentyl)–Br

21. Write the structure of the Grignard reagent formed from the following reactions.

$$C_6H_5\text{-}CH_2Br + Mg \xrightarrow{\text{ether}}$$

$$C_5H_9Br\ (\text{cyclopentyl bromide}) + Mg \xrightarrow{\text{ether}}$$

22. Write the structure of each alcohol that will be obtained by reactions of the formaldehyde (H-CH=O) with Grignard reagents obtained in exercise 21.

8 ALCOHOLS AND PHENOLS

The compounds in which hydroxyl functional group (O-H) is attached to an alkyl chain or a cycloalkyl ring are known as alcohols (**Figure 8.1**). The compounds having hydroxyl group attached to a benzene ring are called phenols.

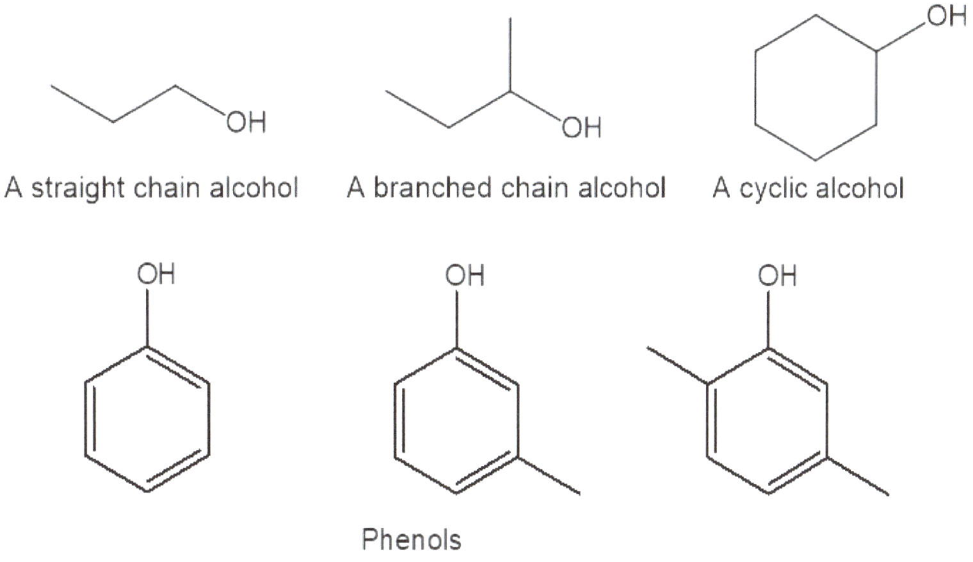

OH
A straight chain alcohol

OH
A branched chain alcohol

OH
A cyclic alcohol

OH OH OH

Phenols

Fig. 8.1: *Some alcohols and phenols.*

The hydroxyl group appears in many biologically important compounds. Bioorganic compounds such as carbohydrates and proteins contain hydroxyl groups. Ethyl alcohol or ethanol is an industrial solvent. It is also used in beverages, and as an additive in fuel. Phenol is commonly known as carbolic acid. It is used as disinfectant. Phenols are also used as antioxidants which our body needs for protection against radicals.

8.1 Nomenclature

Alcohols are named by replacing the last "e" from the name of the parent hydrocarbon with "ol" (**Figure 8.2**). For example, CH_3CH_2OH is called ethanol. Cyclic alcohols are named in a similar manner.

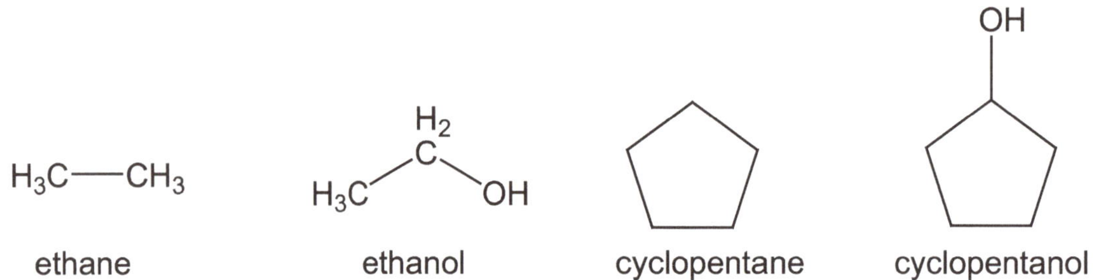

Fig. 8.2: *Nomenclature of simple alcohols.*

If the parent chain has more than two carbon atoms, the hydroxyl group may occupy different positions. In such cases, the position of the hydroxyl group must be shown with an Arabic numeral (**Figure 8.3**). The counting should be done from the side that gives minimum number to the hydroxyl group.

$H_3C—CH_2—CH_2—OH$

1-propanol

$H_3C—CH(OH)—CH_3$

2-propanol

$H_3C—CH(OH)—CH_2—CH_3$

2-butanol
NOT
3-butanol

Fig. 8.3: *Nomenclature in alcohols with more than two carbon atoms.*

The hydroxyl group will have priority over alkyl branches and carbon-carbon double or triple bond in the parent chain that is the hydroxyl group will be given the lowest number (**Figure 8.4**).

$H_3C—CH{=}CH—CH_2—OH$

but-2-en-1-ol

$HC{\equiv}C—CH_2—CH_2—OH$

but-3-yn-1-ol

Fig. 8.4: *Nomenclature of alcohols having carbon-carbon double or triple bond in the parent chain.*

Among phenolic compounds, if a halogen, nitro, alkyl, acyl (RCO) and alkoxy (OR) group is also attached to the benzene ring together with a hydroxyl group the compounds are named as derivatives of phenol (**Figure 8.5**).

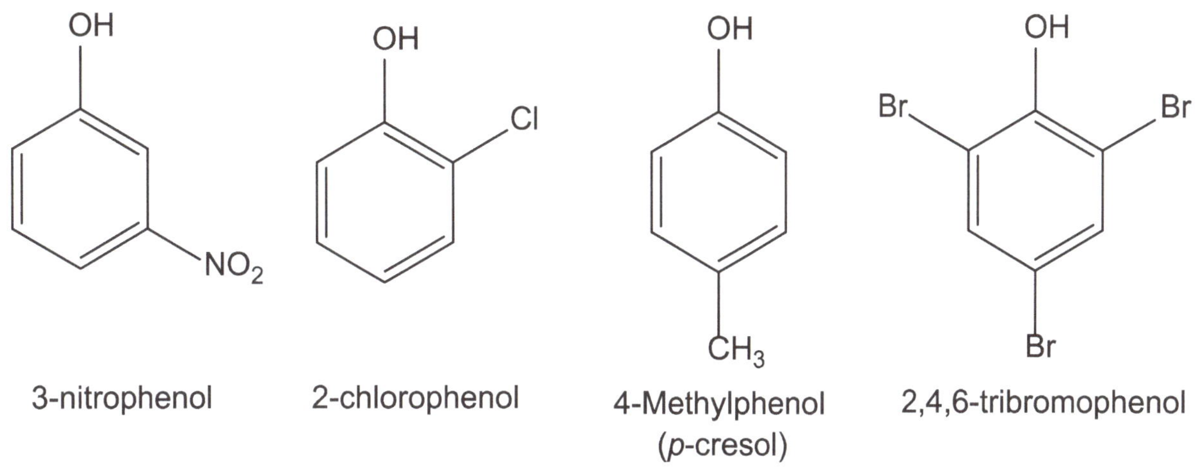

OH, NO_2 — 3-nitrophenol

OH, Cl — 2-chlorophenol

OH, CH_3 — 4-Methylphenol (*p*-cresol)

OH, Br, Br, Br — 2,4,6-tribromophenol

Fig. 8.5: *Nomenclature of phenols.*

Problem: Draw the structures of i) 2-methylcyclopenten-1-ol, ii) hex-3-yn-1-ol, iii) *meta*-bromophenol.

Solution:

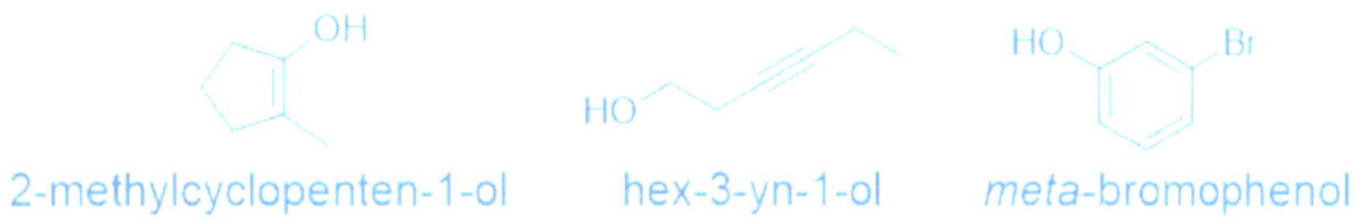

OH — 2-methylcyclopenten-1-ol

HO — hex-3-yn-1-ol

HO, Br — *meta*-bromophenol

8.2 Classification of Alcohols

Alcohols are classified as primary (1°), secondary (2°) and tertiary (3°) just like alkyl halides depending on the carbon atom to which the hydroxyl group is attached (**Figure 8.6**). In primary alcohols, the carbon atom attached to hydroxyl group bears two or more hydrogen atoms whereas in secondary and tertiary alcohols it bears one and no hydrogen atom, respectively.

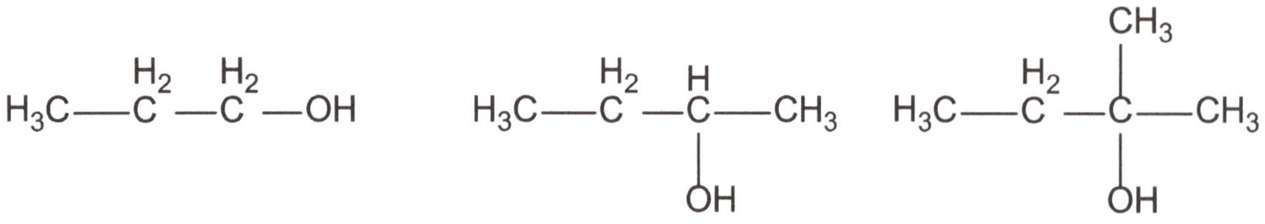

$H_3C-CH_2-CH_2-OH$ 1-Propanol (a primary alcohol)

$H_3C-CH_2-CH(OH)-CH_3$ 2-Propanol (a secondary alcohol)

$H_3C-CH_2-C(CH_3)(OH)-CH_3$ 2-methyl-2-butanol (a tertiary alcohol)

Fig. 8.6: *Classification of alcohols.*

8.3 Preparation of Alcohols and Phenols

There are many methods for preparation of alcohols and phenols. You have studied so far, the hydration of alkenes with water, hydroboration-oxidation sequence of alkenes, and nucleophilic substitution reactions of alkyl halides that lead to the formation of alcohols. The reaction of the Grignard reagents with carbonyl compounds also gives alcohols (see Chapter 10). The preparation of alcohols by reduction of C=O is described in chapter on aldehydes and ketones. A simple method for preparation of phenols is the hydrolysis of sodium salt of benzene sulfonic acid, sodium benzenesulfonate by treating with an acid.

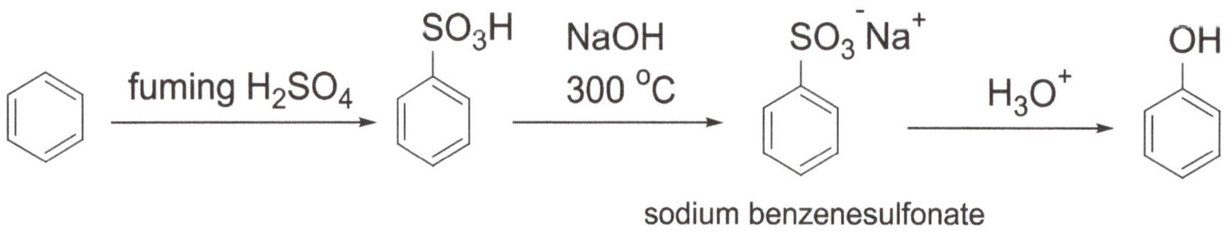

$$C_6H_6 \xrightarrow{\text{fuming } H_2SO_4} C_6H_5SO_3H \xrightarrow[300\ ^oC]{NaOH} C_6H_5SO_3^-Na^+ \xrightarrow{H_3O^+} C_6H_5OH$$

sodium benzenesulfonate

8.4 Physical Properties: Hydrogen Bonding

The boiling points of alcohols are much higher than those of ethers and hydrocarbons of same molecular weight (**Figure 8.7**).

	CH_3CH_2OH	CH_3OCH_3	$CH_3CH_2CH_3$
mol. weight:	46	46	44
b. p. (°C):	78.5	24	- 42

Fig. 8.7: *A comparison of boiling points of ethanol, dimethyl ether, and propane.*

The reason for the high boiling point of alcohols is the ability of its molecules to form hydrogen bond with each other (**Figure 8.8**). Hydrogen bond is a weak bond formed between a hydrogen atom and more electronegative atoms like oxygen, nitrogen, fluorine, etc. The oxygen-hydrogen bond is a highly polar bond because of the big difference in electronegativity of oxygen and hydrogen atoms. As a result, the hydrogen atom carries a partial positive charge whereas the oxygen atom carries a partial negative charge. Thus, the hydrogen atom of one molecule gets attached to the oxygen atom of another molecule by a weak bond called hydrogen bond.

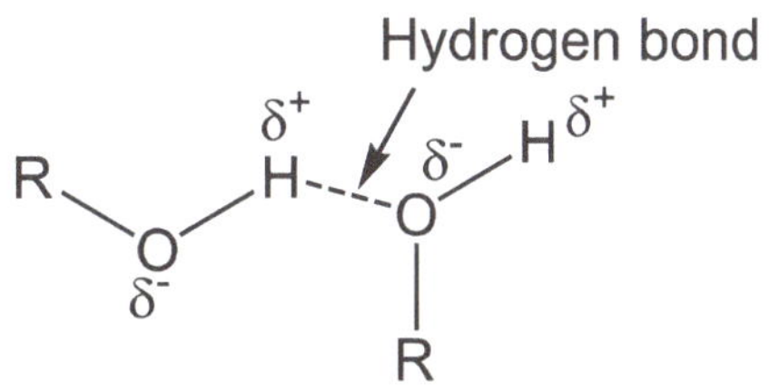

Hydrogen bond

Fig. 8.8: *Hydrogen-bonding in alcohol molecules.*

The lower alcohols (up to propanol) are completely miscible in water. The complete miscibility of alcohols in water is also attributed to the formation of hydrogen bond between water and alcohol molecules. The high molecular weight alcohols do not dissolve in water because with increasing carbon chain length the hydrocarbon character (non-polar) dominates in them.

8.5 Acidity and Basicity of Alcohols and Phenols

i. Alcohol is weakly acidic like water. The dissociation occurs like that in water. Alkoxides, the conjugate bases of alcohols, are strong bases like hydroxide ions. They are ionic compounds and can be easily prepared by reacting sodium with alcohols.

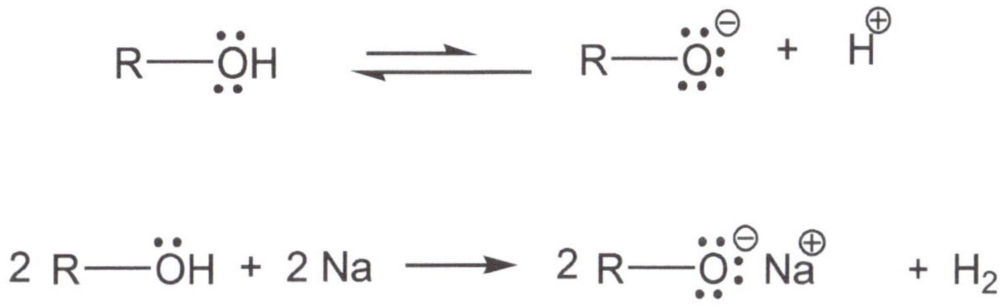

$$R-\ddot{O}H \rightleftharpoons R-\ddot{O}:^{\ominus} + H^{\oplus}$$

$$2\,R-\ddot{O}H + 2\,Na \longrightarrow 2\,R-\ddot{O}:^{\ominus}Na^{\oplus} + H_2$$

ii. Phenols are also weak acids, but they are stronger acids than alcohols mainly because the corresponding phenoxide ion is stabilized by resonance (**Figure 8.9**). The negative charge in alkoxides is localized on oxygen atom.

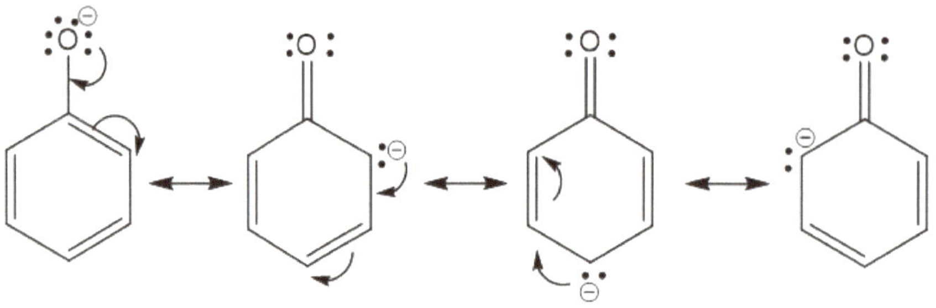

Fig. 8.9: *Resonance-stabilization of phenoxide ion.*

iii. The presence of an electron-withdrawing group on the benzene ring further increases the acidity of phenols by stabilizing the conjugate base (**Figure 8.10**). The electron-donating groups destabilize the conjugate base and so decrease the acidity of these compounds.

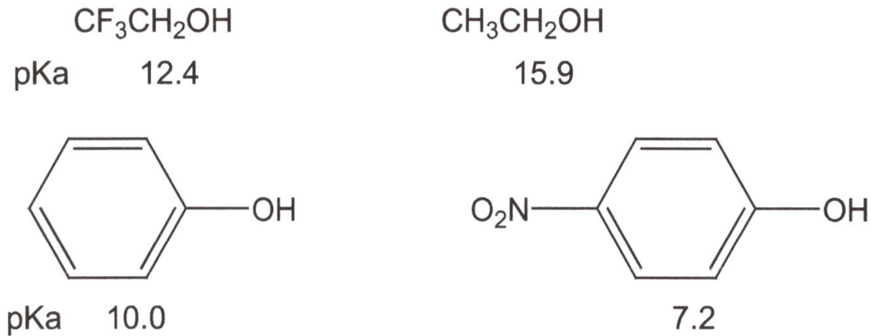

Fig. 8.10: *pKa of some alcohols and phenols.*

iv. Since alcohols and phenols have unshared electron pair on oxygen atom, they can accept a proton and hence these compounds also behave as a weak base.

$$R-\ddot{O}H + H^{\oplus} \rightleftharpoons R-\ddot{O}^{\oplus}H_2$$

Problem: Arrange the following compounds in order of increasing acidity.

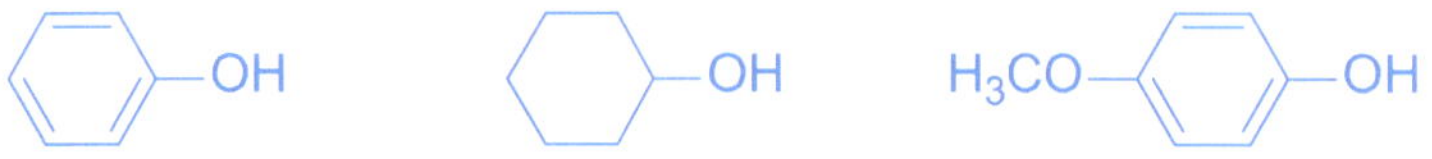

OH   OH H_3CO— —OH

Solution:

Cyclohexanol-OH < H_3CO-C$_6$H$_4$-OH < C$_6$H$_5$-OH

8.6 Chemical Properties of Alcohols

a. Dehydration of Alcohols to Alkenes

The dehydration of alcohols is reverse of the addition of water (hydration) on alkenes. Alcohols can, thus, be converted into alkenes by heating with strong acids at high temperature.

$$CH_3CH_2\text{-}OH \xrightarrow{H^+,\ 180^oC} H_2C{=}CH_2 + H_2O$$

b. Mechanism of Dehydration

The dehydration can take place by both E1 and E2 mechanisms. The formation of 2-methyl-1-propene by dehydration of *tert*-butyl alcohol constitutes a typical example of dehydration which occurs by E1 mechanism. In this mechanism, the hydroxyl group first gets protonated to form hydronium ion (step 1). The latter loses a water molecule which leads to formation of the most stable tertiary carbocation (step 2). Finally, the removal of a proton from a neighboring methyl group leads to the formation of final product alkene (step 3).

$$(CH_3)_3C\text{-}OH + H^+ \xrightarrow{heat} H_2C{=}C(CH_3)_2 + H_2O$$

$$(CH_3)_3C\text{-}OH + H^+ \rightleftharpoons (CH_3)_3C\text{-}\overset{\oplus}{O}H_2 \quad \text{step 1}$$

$$(CH_3)_3C\text{-}\overset{\oplus}{O}H_2 \longrightarrow (CH_3)_3C^{\oplus} + H_2O \quad \text{step 2}$$

$$H\text{-}CH_2\text{-}C^{\oplus}(CH_3)_2 \longrightarrow H_2C{=}C(CH_3)_2 + H^+ \quad \text{step 3}$$

In dehydration of a primary alcohol (E2 mechanism), the two steps after protonation of hydroxyl group – removal of water and removal of a proton occur simultaneously to avoid a primary carbocation which is not so stable.

$$CH_3CH_2\text{-}OH + H^+ \rightleftharpoons CH_3CH_2\text{-}\overset{\oplus}{O}H_2 \quad \text{step 1}$$

$$H\text{-}H_2C\text{-}CH_2\text{-}\overset{\oplus}{O}H_2 \longrightarrow H_2C{=}CH_2 + H_2O + H^+ \quad \text{step 2}$$

There might be two possible alkenes in dehydration of some alcohols. In such a case, the alkene with highly substituted carbon – carbon double bond will be predominant product. In the reaction shown below, 2-methyl-1-butene is a minor product because one carbon atom involved in double bond formation has no substituents; it is attached to two hydrogen atoms. The product 2-methyl-2-butene is a major one because one of the carbon atoms involved in double bond formation has a methyl substituent while the other has two methyl substituents.

$$H_3C\text{-}C(OH)(CH_3)\text{-}CH_2\text{-}CH_3 \xrightarrow[-\,H_2O]{H^+\ \text{heat}} H_2C{=}C(CH_3)\text{-}CH_2\text{-}CH_3 + \underset{\text{(major)}}{H_3C\text{-}C(CH_3){=}CH\text{-}CH_3}$$

Problem: Write all the possible dehydration products from the following alcohols.

a. 3-methyl-3-pentanol, b. 1-ethylcyclopentan-1-ol

Solution:

From a:

and

From b:

and

c. Reaction of Alcohols with Hydrogen halides

The reaction of alcohols with hydrogen halides, H-X (X = Cl, Br, I), gives the corresponding alkyl halides. It is a useful reaction for the preparation of alkyl halides from alcohols. The *tert*-alcohols react fastest while the primary alcohols react slowest.

$$R{-}OH \quad + \quad H{-}X \longrightarrow R{-}X \quad + \quad H{-}OH$$

$$(H_3C)_3C{-}OH \quad + \quad H{-}Cl \xrightarrow[\text{15 min}]{\text{room temp.}} (H_3C)_3C{-}Cl \quad + \quad H{-}OH$$

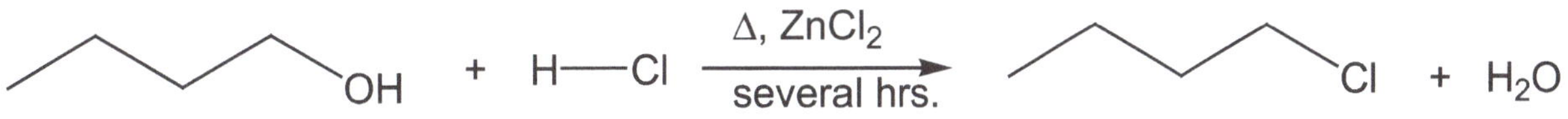

$$CH_3CH_2CH_2CH_2OH \quad + \quad H{-}Cl \xrightarrow[\text{several hrs.}]{\Delta,\ ZnCl_2} CH_3CH_2CH_2CH_2Cl \quad + \quad H_2O$$

The *tert*-alcohols react by S_N1 mechanism whereas the primary alcohols react by S_N2 mechanism. In both the mechanisms, however, the reaction starts with protonation of alcohol forming hydronium ion. In S_N1 mechanism, the hydronium loses water molecule to form carbocation that reacts with halide (chloride in this case) to give the final product. In S_N2 mechanism, Formation of C-Cl bond and loss of water take place simultaneously.

S_N1 mechanism

$$(CH_3)_3C\text{-}OH + H^+ \rightleftharpoons (CH_3)_3C-\overset{\oplus}{O}(H)-H \quad \text{step 1}$$

$$(CH_3)_3C-\overset{\oplus}{O}(H)-H \xrightarrow{\text{slow}} (CH_3)_3C^{\oplus} + H_2O \quad \text{step 2}$$

$$(CH_3)_3C^{\oplus} + Cl^{\ominus} \xrightarrow{\text{fast}} (CH_3)_3CCl \quad \text{step 3}$$

S_N2 mechanism

$$CH_3CH_2CH_2OH + H^+ \rightleftharpoons CH_3CH_2CH_2\overset{\oplus}{O}H_2 \quad \text{step 1}$$

$$Cl^{\ominus} + H_3CH_2C(H)(H)C-\overset{\oplus}{O}(H)-H \longrightarrow CH_3CH_2CH_2Cl + H_2O \quad \text{step 2}$$

d. Reactions of Alcohols with Thionyl chloride and Phosphorus halides

Reactions of alcohols with thionyl chloride and phosphorus halides offer excellent methods for preparation of alkyl halide. These methods are used especially in the case of primary and secondary halides which react very slowly with hydrogen halides.

$$R\text{-}OH \xrightarrow{SOCl_2} R\text{-}Cl$$

$$R\text{-}OH \xrightarrow{PCl_3} R\text{-}Cl$$

e. Oxidation of Alcohols to Aldehydes, Ketones and Carboxylic acids

Oxidation of primary and secondary alcohols using Jone's reagent (chromium trioxide in aqueous sulfuric acid and acetone) forms aldehydes and ketones, respectively. The aldehyde formed by the oxidation of a primary alcohol, however, gets further oxidized to carboxylic acid. Aldehydes can be prepared by oxidation of primary alcohols with a reagent called pyridinium chlorochromate (PCC).

$$R{-}CH_2{-}OH \xrightarrow[\text{acetone}]{CrO_3,\ H^+} [R{-}CH{=}O] \xrightarrow[\text{acetone}]{CrO_3,\ H^+} R{-}C({=}O){-}OH$$

$$R{-}CH(R){-}OH \xrightarrow[\text{acetone}]{CrO_3,\ H^+} R_2C{=}O$$

$$R{-}CH_2{-}OH \xrightarrow[CH_2Cl_2]{PCC} R{-}CH{=}O$$

Problem: Write the chemical equations corresponding to the reaction of i) cyclohexanol with CrO_3 in acetone (solvent), ii) 1-pentanol with CrO_3 in acetone, iii) 2-methylbutanol with PCC in dichloromethane (CH_2Cl_2) as a solvent.

Solution:

i) cyclohexanol $\xrightarrow[\text{acetone}]{CrO_3}$ cyclohexanone ii) 1-pentanol (OH) $\xrightarrow[\text{acetone}]{CrO_3}$ pentanoic acid (OH, O)

ii) 2-methylbutanol (OH) $\xrightarrow[CH_2Cl_2]{PCC}$ 2-methylbutanal (H, O)

8.7 Chemical Properties of Phenols

a. Oxidation of Phenols

Phenols are easily oxidized to colored oxidation products on exposure to air for some time. For example, benzene-1,4-diol (commonly known as hydroquinone) is oxidized to 1,4-benzoquinone (*para*-benzoquinone) using acidic sodium dichromate.

$$\text{HO–}C_6H_4\text{–OH} \xrightarrow[H_2SO_4,\ 30^{o}C]{Na_2Cr_2O_7} O{=}C_6H_4{=}O$$

Hydroquinone (colorless) — 1,4-Benzoquinone (yellow)

b. Electrophilic Aromatic Substitution in Phenols

Phenols undergo electrophilic aromatic substitution readily because the hydroxyl group is a highly ring-activating group. For example, phenol is nitrated with dilute nitric acid to form 4-nitrophenol.

$$\text{HO–}C_6H_5 \xrightarrow{\text{dil } HNO_3} \text{HO–}C_6H_4\text{–}NO_2$$

phenol — 4-nitrophenol

Phenolisalsoreadilybrominatedwithbrominewatertoform2,4,6-tribromophenol.

$$\text{HO–}C_6H_5 \xrightarrow[H_2O]{\text{excess } Br_2} \text{HO–}C_6H_2Br_3$$

phenol — 2,4,6-tribromophenol

SUMMARY

Alcohols are compounds containing an –OH group bonded to an alkyl group. The compounds are called phenols if the –OH group is bonded to a benzene ring. Alcohols can be prepared by many methods including hydration and hydroboration-oxidation of alkenes, nucleophilic substitution of alkyl halides, and reactions of the Grignard reagents with carbonyl compounds.

Alcohols have higher solubility in water than hydrocarbons due to their ability to form hydrogen bonds. Their boiling points are also usually high because they can form intramolecular hydrogen bonds. Alcohols are weakly acidic and are converted to their alkoxides on treatment with a strong base or an alkali metal. Phenols are more acidic in comparison to alcohols because the phenoxide ions are resonance stabilized.

Dehydration to Alkenes

$$\text{H–C–C–OH} \xrightarrow{H^+ \text{(cat.)}} \text{C=C}$$

Conversion to Alkyl halides

$$R\text{—}OH + HX \longrightarrow R\text{—}X + H_2O \quad (X = Cl, Br, I)$$

$$R\text{—}OH + SOCl_2 \longrightarrow R\text{—}Cl + HCl + SO_2$$

$$3\ R\text{—}OH + PX_3 \longrightarrow 3\ R\text{—}X + H_3PO_3 \quad (X = Cl, Br)$$

Oxidation of alcohols to Aldehydes, Ketones and Carboxylic acids

$$R\text{—}CH_2\text{—}OH \text{ (pr. alcohol)} \xrightarrow{PCC} R\text{—}CH\text{=}O$$

$$R\text{—}CH_2\text{—}OH \xrightarrow[H^+]{CrO_3} R\text{—}C(=O)\text{—}OH$$

$$R_2CH\text{—}OH \text{ (sec. alcohol)} \xrightarrow[CrO_3/H^+]{PCC \text{ or}} R_2C\text{=}O$$

Formation of Phenoxide

$$C_6H_5\text{—}OH + NaOH \longrightarrow C_6H_5\text{—}O^-Na^+$$

Electrophilic Aromatic Substitution of Phenol

$$\text{C}_6\text{H}_5\text{OH} \xrightarrow{\text{dil } HNO_3} \text{HO–C}_6\text{H}_4\text{–}NO_2$$

$$\text{C}_6\text{H}_5\text{OH} \xrightarrow[H_2O]{\text{excess } Br_2} \text{2,4,6-tribromophenol}$$

Oxidation of Phenols to Quinones

$$\text{HO–C}_6\text{H}_4\text{–OH} \xrightarrow[H_2SO_4,\ 30^\circ C]{Na_2Cr_2O_7} \text{O=C}_6\text{H}_4\text{=O}$$

Hydroquinone (colorless) → 1,4-Benzoquinone (yellow)

EXERCISES

1. Name the following alcohols according to the IUPAC rules.

   a. $CH_3(CH_2)_6OH$ b. $CH_3CH_2CH(OH)CH_2CH_3$ c. $H_2C{=}CHCH(OH)CH_3$,
   d. $(CH_3)_2CHOH$ e. $CH_3CH(Cl)CH_2CH(OH)CH_3$ f. $(CH_3)_3COH$

2. Write structural formula for each of the following compounds.

   a. 2,2-dimethyl-1-pentanol b. 3-chlorophenol c. 1-bromocyclohexanol
   d. 2-cyclopentenol e. 2-bromo-2-propen-1-ol f. pent-4-yn-1-ol

3. Identify the following alcohols as primary, secondary, or tertiary alcohol.

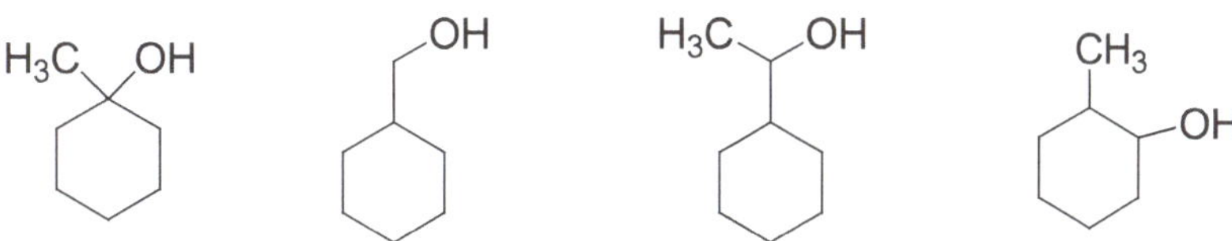

4. Explain the following observations.

   a. Benzyl alcohol, $C_6H_5CH_2OH$, boils at much higher temperature (203 °C) than toluene, $C_6H_5CH_3$, 110 °C.

   b. Alcohols have more solubility than hydrocarbons in water.

   c. The pK_a values of alcohols range from 15 – 18 but phenol has a pKa of 10.

5. Which of the following will be more acidic in each pair and why?

   a. Phenol and 4-ethylphenol

   b. 4-methylphenol and 4-chlorophenol

6. Show the most appropriate equation for the preparation of following alcohols from alkenes.

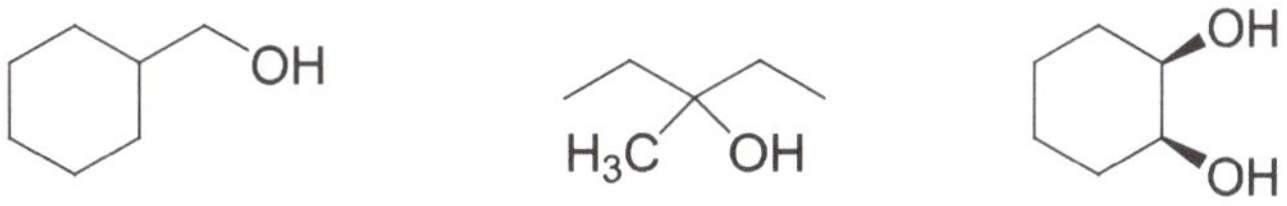

OH

H_3C OH

OH

OH

7. Show the reaction scheme for preparation of following phenols starting from benzene.

   a. 3-methylphenol b. 4-chlorophenol c. 3-nitrophenol

8. Show the structure of product obtained from the following reactions of alcohols and phenols.

a.  OH + Na →

c.  O_2N–⟨benzene⟩–OH + KOH →

b.  OH + NaOH →

d.  –OH + K →

9. Write chemical equations showing formation of all possible acid-catalyzed dehydration products of the following alcohols. If more than one alkene is possible, predict the major one.

   a. 2-pentanol b. cyclohexanol c. 1-methylcyclobutanol

   d. 2-phenylethanol e. 2-methyl-2-pentanol

10. Write out all the steps in the mechanism of acid-catalyzed dehydration of 2-methyl-2-butanol.

11. What alcohols can be used for the preparation of following alkenes by dehydration?

a. $H_2CH_2CHC{=}CHCH_2CH_2$ b. 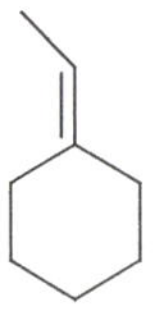 c. 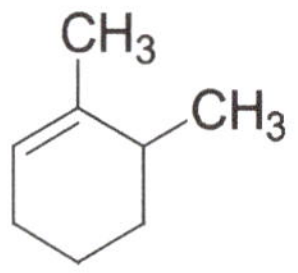 CH_3 CH_3

12. Write the product of the following reactions.

a. 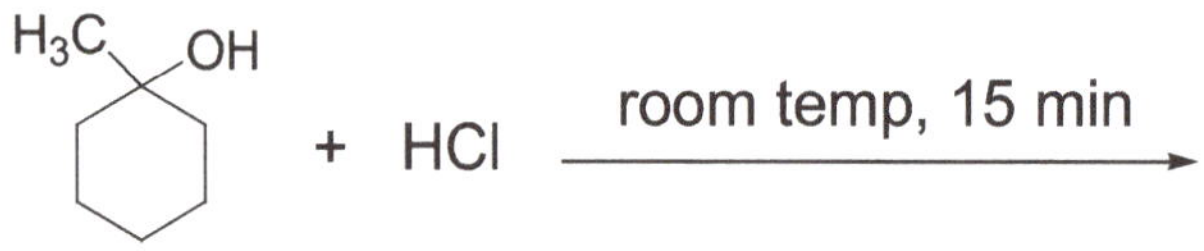 H_3C OH + HCl $\xrightarrow{\text{room temp, 15 min}}$

b.  OH + $SOCl_2$ $\xrightarrow{\text{heat}}$

c. 3 $CH_3CH_2CH(OH)CH_2CH_3$ + PCl_3 $\longrightarrow$

d. 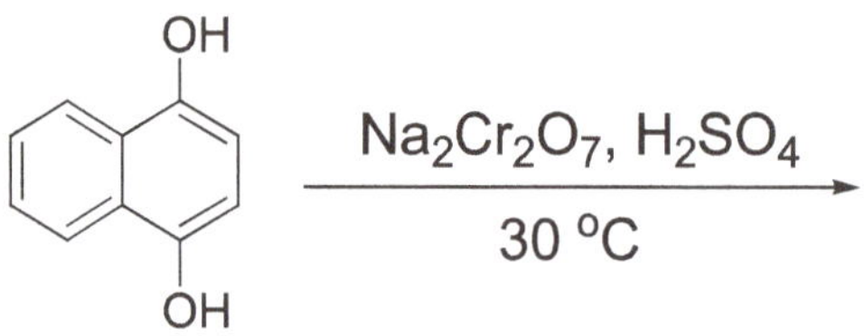 OH OH $\xrightarrow[\text{30 °C}]{Na_2Cr_2O_7,\ H_2SO_4}$

13. Write an equation for the oxidation of each of following alcohols using the reagents provided.

    a. 1-butanol with Jone's reagent.

    b. 1-hexanol with PCC.

    c. 5-phenyl-2-pentanol with PCC.

    d. 5-phenyl-2-pentanol with Jone's reagent.

    e. 2-methylcyclohexanol with Jone's reagent.

14. Complete the following reaction chart by providing the missing reagent or product.

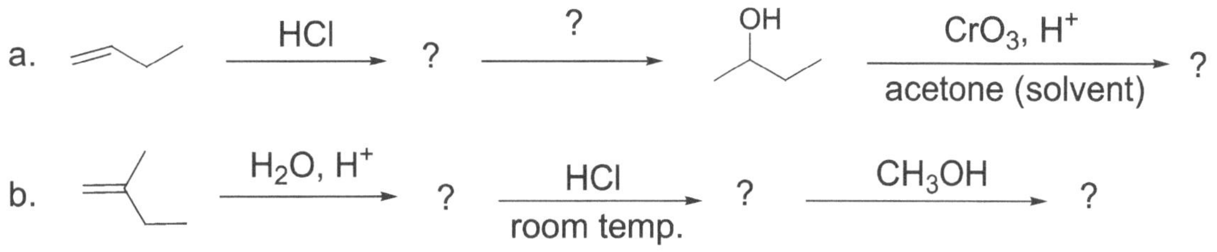

a. HCl → ? —?→ OH —CrO_3, H^+ / acetone (solvent)→ ?

b. H_2O, H^+ → ? —HCl / room temp.→ ? —CH_3OH→ ?

9 ETHERS AND EPOXIDES

Ethers are compounds that have at least one oxygen atom attached to two organic groups R-O-R', where R and R' may be same or different: alkyl or aromatic. The word ether, for many people, is associated with the well-known general anesthetic. Ether is also a common ingredient starter fluid for car engines. Methyl phenyl ether, commonly known as anisole, is a constituent of the essential oil of anise seed.

9.1 Nomenclature

Ethers are named by putting the names of the organic groups as branches in alphabetical order, followed by the word ether (**Figure 9.1**).

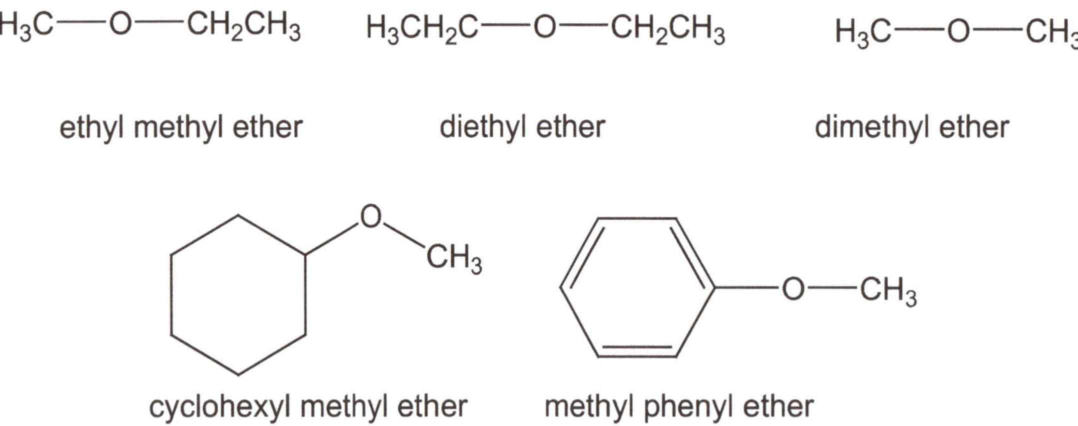

Fig. 9.1: *Naming simple ethers.*

For others with complex structures, -OR group is named as an alkoxy group (**Figure 9.2**).

$—OCH_3$	$—OC_2H_5$	$—OCH_2CH_2CH_3$	$—OCH(CH_3)_2$	$—OC(CH_3)_3$
methoxy	ethoxy	propoxy	isopropoxy	*tert*-butoxy

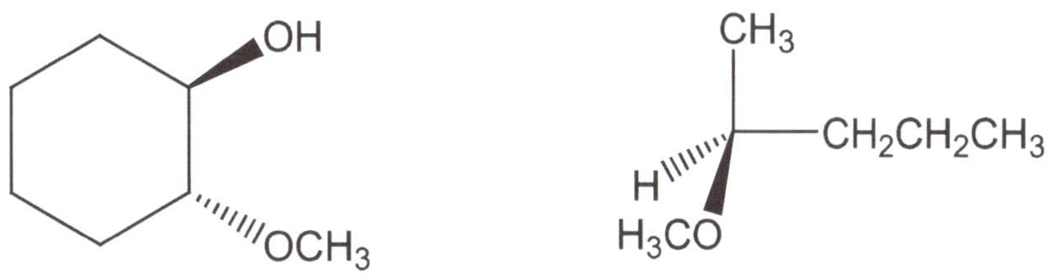

(1*R*,2*R*)-2-methoxycyclohexanol

(*S*)-2-methoxypentane

Fig. 9.2: *Naming ethers as alkoxides.*

9.2 Preparation of Ethers

a. Commercial Production of Ethers

The most important commercial ether, diethyl ether, is prepared by heating ethanol with sulfuric acid at 140 °C. The temperature control is important in this method because ethanol is dehydrated to ethylene at 180 °C.

$$CH_3CH_2OH + CH_3CH_2OH \xrightarrow[140\ ^{o}C]{H_2SO_4} CH_3CH_2OCH_2CH_3$$

tert-Butyl methyl ether, used in gasoline to increase its *Octane Rating* is commercially prepared by acid-catalyzed addition of methanol on 2-methylpropene.

$$CH_3OH + H_2C{=}C(CH_3)_2 \xrightarrow{H^+} CH_3OC(CH_3)_3$$

b. Williamson Synthesis of Ethers

The Williamson synthesis is the most important method for preparation of ethers in laboratory. This method involves preparation of an alkoxide by reaction of an alcohol with sodium metal. The sodium alkoxide is then reacted with an alkyl halide. Since the reaction of alkyl halides with alkoxides takes place by an S_N2 mechanism, the method works best if the halide is primary one.

$$R'—X + R—\overset{\ominus}{O}Na^{\oplus} \longrightarrow R'—O—R + NaX$$

Problem: Write equations for the synthesis of the following ethers by the Williamson method.

i) C_6H_5—OCH_2CH_3 ii) $(CH_3)_3COCH_2CH_3$

Solution: Take primary alkyl halides and appropriate alkoxide or phenoxide.

i) $2\ C_6H_5\text{—}OH \xrightarrow{2\ Na} 2\ C_6H_5\text{—}O^-Na^+$

$C_6H_5\text{—}O^-Na^+ \xrightarrow{CH_3CH_2Br} C_6H_5\text{—}OCH_2CH_3 + NaBr$

ii) $(CH_3)_3COH \xrightarrow{2\ Na} 2\ (CH_3)_3CO^-Na^+$

$(CH_3)_3CO^-Na^+ + CH_3CH_2Br \longrightarrow (CH_3)_3COCH_2CH_3 + NaBr$

9.3 Physical Properties

Ethers are colorless compounds with relatively pleasant odor. They boil at much lower temperatures than alcohols with an equal number of carbon atoms. It is because ether molecules cannot form hydrogen bonds with one another. Ether molecules, however, can form hydrogen bonds with alcohol molecules (**Figure 9.3**). As a result, ether and alcohol are mutually soluble liquids.

R
O----H
R O
R

Fig. 9.3: *Hydrogen-bonding of ether with alcohol.*

a. Ethers as Solvents

Ethers are excellent solvents for carrying out organic reactions. Most of the organic compounds are soluble in ether. Furthermore, ethers are relatively inert compounds. They do not react with dilute acids, with dilute bases or with common oxidizing and reducing agents. Ethers are also used for extraction of organic compounds from natural sources.

b. Role of Ethers in Formation of the Grignard Reagents

The preparation of the Grignard reagents is a nice example of the application of ethers as a solvent. The most common Grignard reagents are prepared by the reaction of alkyl or aromatic halides with magnesium in diethyl ether. A cyclic ether tetrahydrofuran (THF) is also frequently used as a solvent in preparation of Grignard reagents.

$$R{-}X + Mg \xrightarrow[0\ ^{o}C]{\text{dry ether}} R{-}Mg{-}X$$

The ether does not work only as a solvent but the lone pair on oxygen coordinates with magnesium in the Grignard reagent and stabilizes it. The magnesium in Grignard reagent is two electron pairs short of an octet. The oxygen of an ether molecule can donate an electron pair to the magnesium (**Figure 9.4**).

C_2H_5–Ö–C_2H_5 → R–Mg–Br ← C_2H_5–Ö–C_2H_5

Fig. 9.4: *Coordination of oxygen in ether with magnesium of Grignard reagent.*

9.4 Cleavage of Ethers

Ethers react with strong protic acids and with Lewis acids. Ethers are Lewis bases because the oxygen atom in it has unshared electron pairs.

$$R{-}\ddot{O}{-}R' + H^{+} \rightleftharpoons R{-}\overset{H}{\overset{|}{O^{\oplus}}}{-}R'$$

$$R-\ddot{O}-R' + Br_3B \rightleftharpoons Br_3B^{\ominus}-\ddot{O}^{\oplus}(R)(R')$$

The above reactions are like the reaction of alcohols with strong acids. If the alkyl groups R and/or R′ are primary or secondary, the bond to oxygen can be broken by reacting ethers with a strong nucleophile such as Br^- or I^- (by S_N2 mechanism). For example,

$$H_3CH_2C-O-CH(CH_3)_2 + HI \xrightarrow{\text{heat}} H_3CH_2C-I + HO-CH(CH_3)_2$$

$$C_6H_5-O-CH_3 + Br_3B \xrightarrow[H_2O]{\text{heat}} C_6H_5-OH + CH_3Br$$

If both the organic groups attached to oxygen atom are tertiary, the reaction takes place by an S_N1 mechanism and hence no strong nucleophile is required.

$$C_6H_5-O-C(CH_3)_3 \xrightarrow[H_2O]{H^+} C_6H_5-OH + (CH_3)_3C\text{-}OH$$

Always keep in mind that in S_N2 reaction the nucleophile will react at the less hindered site and in S_N1 reaction the formation of most stable carbocation will influence the cleavage of the C-O-C bond.

Problem: Write the products and stepwise mechanism of their formation when *n*-butyl isopropyl ether is reacted with HBr.

Solution: Since the oxygen atom is attached to a primary and a secondary alkyl group and the acid is a strong one, the S_N2 mechanism will be preferred and bond between *n*-butyl group and ether will be cleaved forming *n*-butyl bromide and isopropyl alcohol.

$$H_3CH_2CH_2CH_2C-\ddot{O}-\overset{H}{C}(CH_3)-CH_3 + H^+ \rightleftharpoons H_3CH_2CH_2CH_2C-\overset{H}{O^{\oplus}}-\overset{H}{C}(CH_3)-CH_3$$

$$\xrightarrow{Br^{\ominus}} H_3CH_2CH_2CH_2C-Br + (H_3C)_2CH-\ddot{O}H$$

9.5 Cyclic Ethers: Epoxides

There are many ethers known in which oxygen atom is present in the ring (cycle) (**Figure 9.5**). The smallest possible cyclic ether is known as epoxide or oxirane (oxacyclopropane). Epoxides are very reactive compounds because three-membered ring is highly strained. The ring easily cleaves forming open chain products. A cyclic ether named tetrahydrofuran (oxacyclopentane), abbreviated as THF, is an excellent solvent for chemical reactions. 1,4-Dioxane having two oxygen atoms in ring is also a useful solvent. Macrocyclic polyethers, crown ethers are used as phase-transfer catalysts.

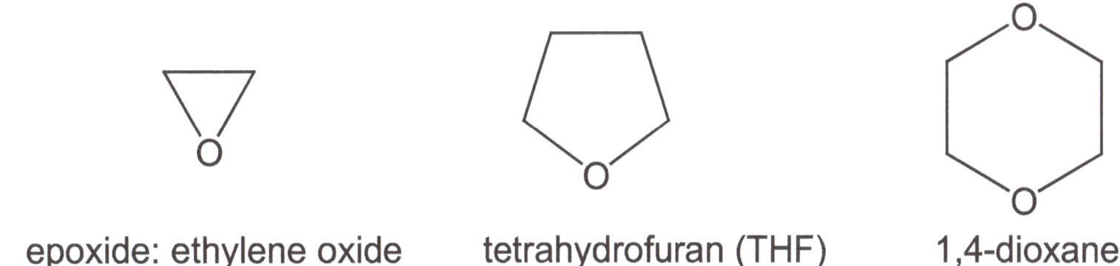

Fig. 9.5: *Some common cyclic ethers.*

Although many cyclic ethers are known with their common names you can name them as oxacycloalkane by counting the total number of atoms in the ring (**Figure 9.6**). To locate the position of substituent(s) on the ring, always start numbering from oxygen atom in the direction giving minimum number to the branch.

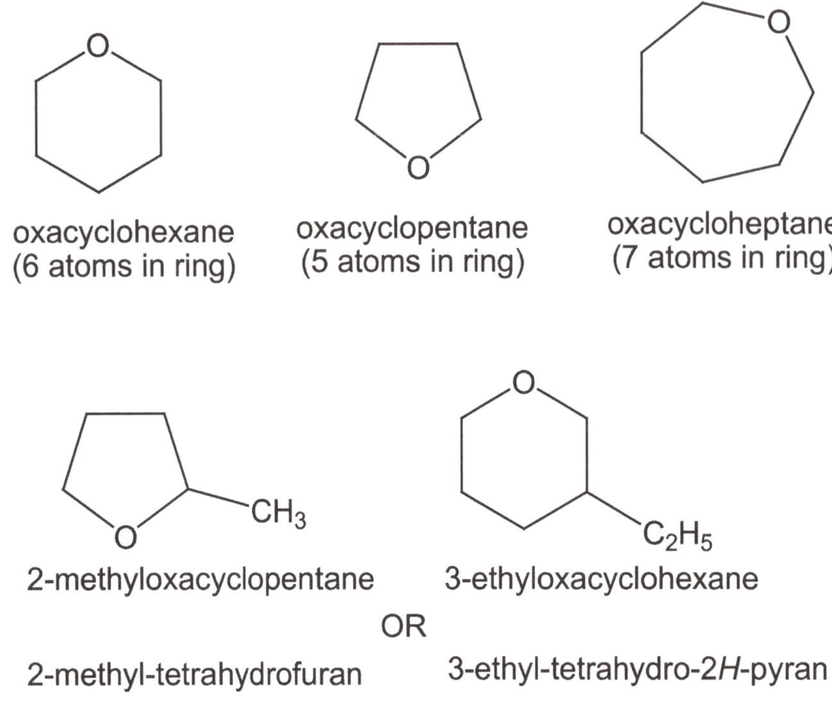

Fig. 9.6: *Naming cyclic ethers.*

a. Preparation of Cyclic Ethers

Ethylene oxide is the commercially most important epoxide. It is prepared by silver-catalyzed air oxidation of ethylene. It is a useful raw material for the preparation of ethylene glycol. Ethylene glycol is used in automobile cooling systems as antifreeze. It is also used in preparation of the polyester fibers.

$$H_2C{=}CH_2 + O_2 \xrightarrow[250\ ^{\circ}C,\ \text{pressure}]{\text{Ag catalyst}} \text{ethylene oxide}$$

Other epoxides are usually prepared by oxidation of an alkene with an organic peracid.

$$\text{cyclohexene} + R{-}\overset{\overset{\displaystyle O}{\|}}{C}{-}O{-}O{-}H \longrightarrow \text{cyclohexene oxide} + R\text{-}COOH$$

b. Cleavage of Cyclic Ethers

Nucleophiles react at the carbon atoms of the epoxides leading to the opening of the ring. For example, the reaction of ethylene oxide with water in the presence of an acid catalyst leads to the formation of ethylene glycol (ethane-1,2-diol). The Grignard reagents also react with epoxides to afford alcohols. The reaction of alcohols with the epoxides will form compounds containing one ether and one alcohol functional group.

ethylene oxide + H—OH $\xrightarrow{H^+}$ HO–CH₂CH₂–OH

ethylene glycol (a 1,2-diol)

Epoxide $\xrightarrow[2.\ H_2O]{1.\ R\text{-}MgX}$ R–CH₂CH₂–OH

Epoxide + CH_3OH $\xrightarrow{H^+}$ H_3CO–CH₂CH₂–OH

In the presence of acid, the reaction proceeds through protonation of the oxygen atom in epoxide ring followed by nucleophilic attack on ring carbon. In the absence of acid, the nucleophiles attack the ring carbon which is followed by ring-cleavage. The reaction of ethylene oxide with Grignard reagents is an important method for preparing open-chain alcohols with desired number of carbon atoms in the chain. Aromatic and other cyclic Grignard reagents can attach aromatic or other rings on to the alcohol chain.

Epoxide $\underset{}{\overset{H^+}{\rightleftharpoons}}$ protonated epoxide (⊕O–H) $\xrightarrow{H_2O}$ HO–CH₂CH₂–⊕O(H)–H $\rightleftharpoons$ $HO\text{-}CH_2CH_2\text{-}OH + H^+$

Epoxide + $\overset{\delta -}{R}$—$\overset{\delta +}{MgX}$ → XMgO–CH₂CH₂–R $\xrightarrow{HOH}$ HO–CH₂CH₂–R + Mg(OH)X

SUMMARY

Ethers (R-O-R) are important solvents. Ethers play a crucial role in the formation of extremely valuable Grignard reagents. Non-bonded electron pairs on ether molecules coordinate with magnesium metal to provide stability to Grignard reagents.

The most important method for preparation of ethers is the Williamson synthesis in which alcohols are reacted with sodium metal to get the corresponding alkoxides which then react as nucleophile with alkyl halides to give the ethers. Ethers are also prepared by catalytic addition of alcohols to alkenes. Cyclic ethers, epoxides are prepared by oxidation of alkenes with peroxy acids; and can be converted to open-chain ethers by reaction with alcohols under acidic condition.

Cleavage of Ethers by Hydrogen Halides and BBr_3

$$R{-}O{-}R + H\text{-}X \longrightarrow R{-}OH + R{-}X$$

$$R{-}O{-}R \xrightarrow[2.\ H_2O]{1.\ BBr_3} R{-}OH + R{-}Br$$

Cleavage of Epoxide by Water and Alcohols

$$\text{epoxide} \xrightarrow{H^+/H\text{-}OH} HO{-}CH_2CH_2{-}OH$$

$$\text{epoxide} \xrightarrow{H^+/R\text{-}OH} HO{-}CH_2CH_2{-}OR$$

Cleavage of Epoxide by Grignard Reagents

$$\text{epoxide} + \overset{\delta -}{H_3C}{-}\overset{\delta +}{MgX} \longrightarrow XMgO{-}CH_2CH_2{-}CH_3 \xrightarrow{HOH} HO{-}CH_2CH_2{-}CH_3 + Mg(OH)X$$

EXERCISES

1. Write a structural formula for each of the following compounds containing ether linkage.

   a. methyl phenyl ether b. diisopropyl ether c. diphenyl ether

   d. benzyl phenyl ether e. 2-methoxybut-1-ene f. propylene oxide

   g. *cis*-3-methoxycyclohexanol h. *p*-ethoxyphenol i. 1,3-dioxane

2. Name each of the following compounds.

   a. $CH_3CH_2OC(CH_3)_3$ b. $CH_3CH_2CH_2OCH(CH_3)_2$ c. $H_2C{=}CHCH_2CH_2OCH_3$

   d.  e. 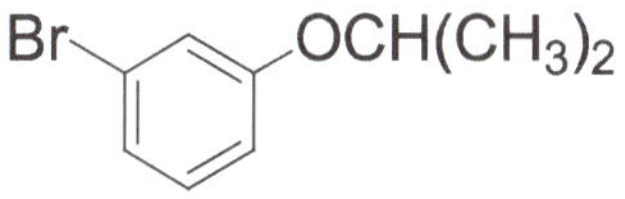 Br, $OCH(CH_3)_2$ f. 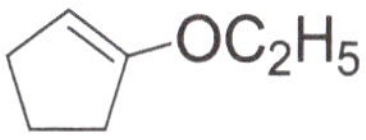 OC_2H_5 g. 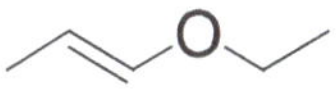 O

3. Write all possible isomers of alcohols and ethers with molecular formula C_3H_7O.

4. Arrange the following compounds in order of decreasing solubility in water.

   a. $CH_3(CH_2)_4CH_3$ b. $HOH_2CCH_2CH_2OH$ c. $CH_3CH_2OCH_3$ d. $CH_3CH_2CH_2OH$

5. Briefly describe the role of ethers in the formation of Grignard reagents.

6. Describe at least three methods for the preparation of ethers. Write an equation corresponding to each method. Show equation for preparation of ethyl propyl ether by the Williamson's method.

7. Write equations for the synthesis of the following ethers by the Williamson's method.

   a. ethoxybenzene b. *tert*-butyl ethyl ether c. methoxycyclohexane

8. What products would you expect from the reaction of the following ethers with HBr?

   a. $CH_3CH_2CH_2OCH_2CH_2CH_3$ b. 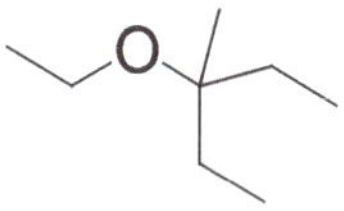 O c. 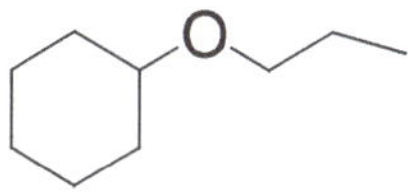 O

9. Write the structure of the products when ethyl phenyl ether is heated with boron tribromide and then treated with water.

10. Show equation for the preparation of phenyl epoxide using an appropriate alkene and a peroxy acid.

11. Complete the following reaction scheme by providing the missing product/ reagent.

CO_3H
Cl
?
?
OC_2H_5
OH
H^+
H_2O
?

12. How can you prepare 2-phenylethanol from ethylene oxide? Show equation.

13. A cyclic ether reacts with water in the presence of a protic acid to give butane-1,4-diol. Show the structure of the cyclic ether.

14. Show stepwise mechanism of the following reaction.

O
H^+
H_2O
OH
OH

10 ALDEHYDES AND KETONES

Aldehydes and ketones constitute a large class of organic compounds which touch everyday life. Cinnamaldehyde which occurs in cinnamon is responsible for the flavor of cinnamon. Formaldehyde is a component of many building materials. Civetone, a ketone is used to get musky odor in many perfumes. Many hormones in the body which are responsible for biological processes belong to the class of aldehydes and ketones.

Aldehydes and ketones contain carbonyl group, C=O, as the functional group. In the aldehyde functional group, the carbon of the carbon-oxygen double bond must be attached to at least one hydrogen atom (-CHO) whereas in ketones it is attached to two carbon atoms (**Figure 10.1**).

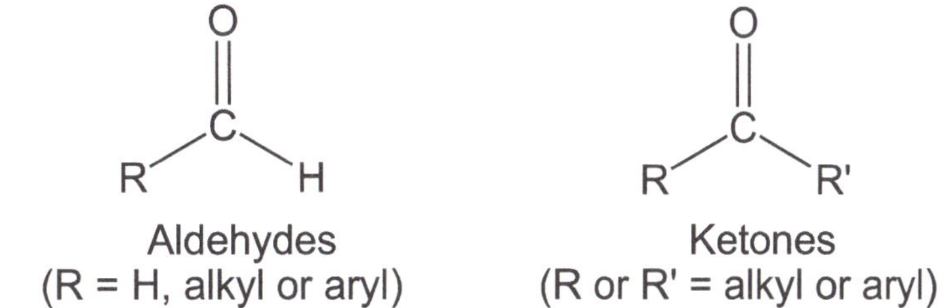

Fig. 10.1: *General structural formula of aldehydes and ketones*

10.1 Some Common Aldehydes and Ketones

Formaldehyde, a gas supplied as 37% aqueous solution, is produced by catalytic dehydrogenation of methanol in the presence of silver catalysts at 600-700 °C. It is used in the manufacture of plastics, building insulation, plywood, etc.

$$CH_3OH \xrightarrow[600\text{-}700\ ^oC]{\text{Ag catalyst}} H_2C{=}O + H_2\uparrow$$

Acetaldehyde, used for the preparation of acetic acid and other chemicals, is prepared by oxidation of ethylene in the presence of Pd-Cu catalyst at 100-300 °C.

$$2\ H_2C{=}CH_2 \xrightarrow[100\text{-}300\ ^\circ C]{Pd\text{-}Cu} 2\ H_3CHC{=}O$$

The above method is also used for production of acetone. However, acetone can also be prepared by oxidation of isopropyl alcohol using Jone's reagent or PCC.

$$H_3C{-}CH{=}CH_2 \xrightarrow[100\text{-}300\ ^\circ C]{Pd\text{-}Cu} (H_3C)_2C{=}O$$

10.2 Nomenclature of Aldehydes and Ketones

In the IUPAC system, aldehydes are named by replacing the terminal "e" of the alkyl chain with "al". The parent chain must contain the –CHO functional group. The numbering of the chain should start from the carbon of the functional group (**Figure 10.2**).

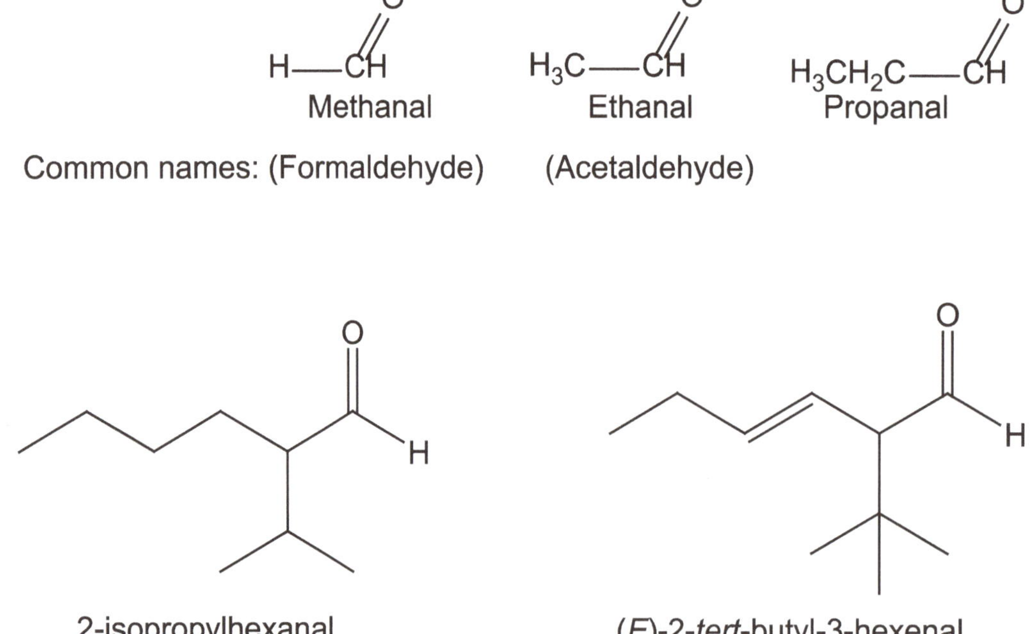

Fig. 10.2: *Naming open-chain aldehydes.*

In cyclic aldehydes, the suffix carbaldehyde is used (**Figure 10.3**). However, for a –CHO attached to benezene ring the common name benzaldehyde is used.

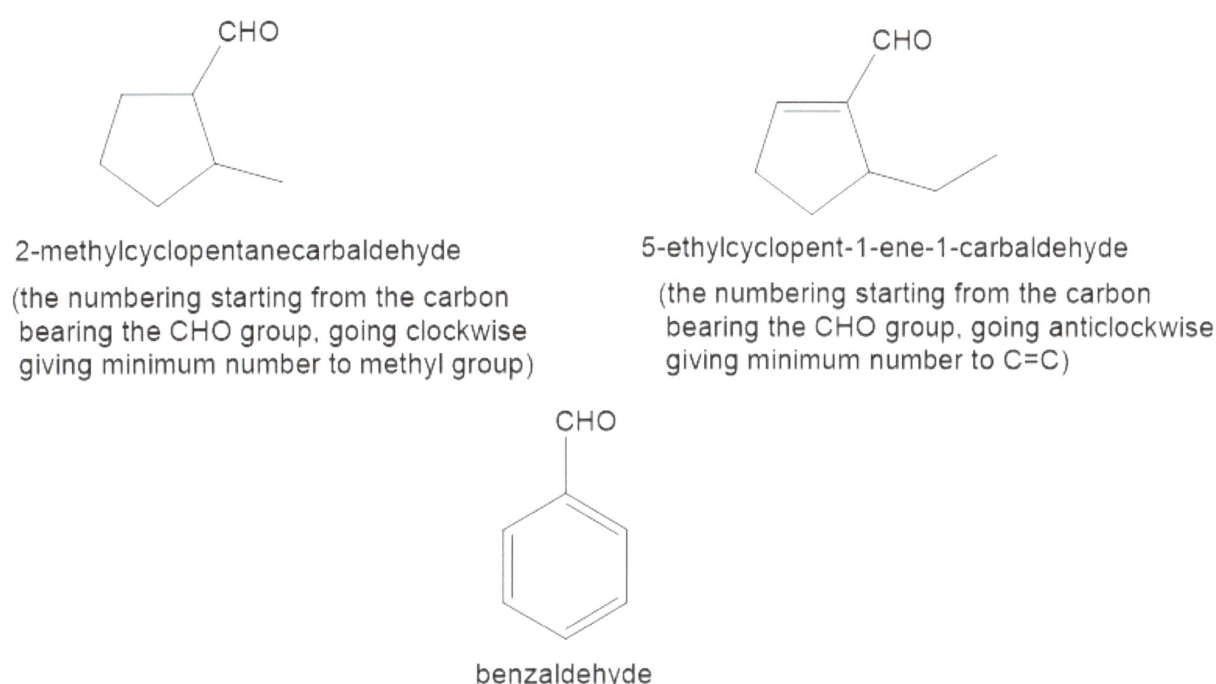

Fig. 10.3: *Naming cyclic aldehydes.*

Ketones are named by replacing the terminal "e" of the alkyl chain with "one" (**Figure 10.4**). The parent chain must contain the –C=O functional group and numbering of the chain should start from the side that gives minimum number to the carbonyl group. However, many ketones have common names and many of them are named by writing the names of the organic groups attached to carbonyl groups followed by word "ketone". Some examples of nomenclature in ketones are shown below.

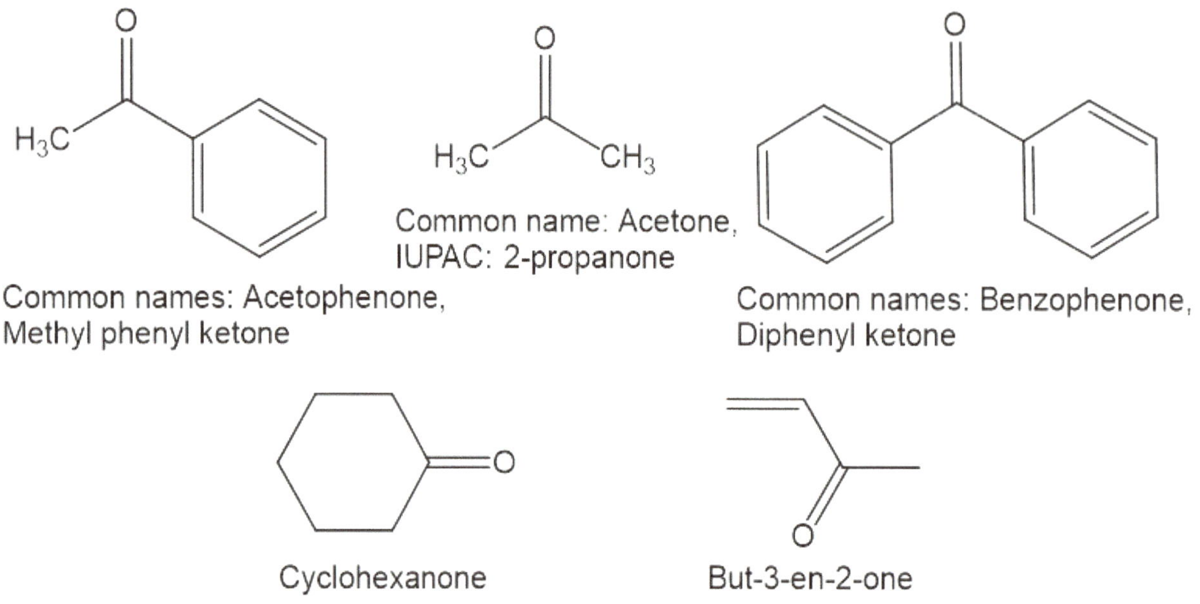

Common names: Acetophenone, Methyl phenyl ketone

Common name: Acetone, IUPAC: 2-propanone

Common names: Benzophenone, Diphenyl ketone

Cyclohexanone

But-3-en-2-one

Fig. 10.4: *Naming ketones.*

The R-C=O groups are also named as acyl or benzoyl groups. For example, $-COCH_3$ group is called acetyl and COC_6H_5 group is called benzoyl (**Figure 10.5**). See the following examples of carboxylic acids.

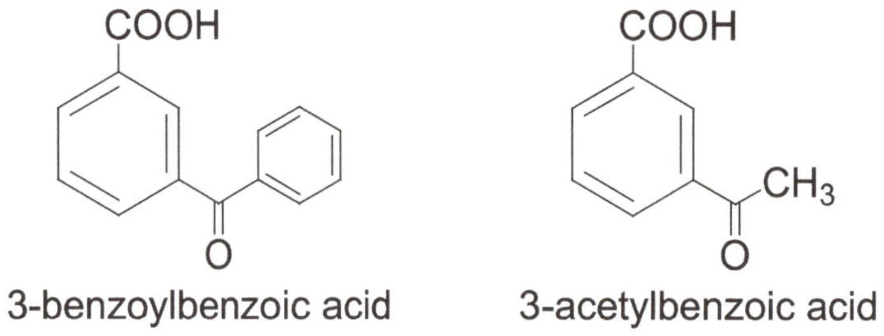

3-benzoylbenzoic acid

3-acetylbenzoic acid

Fig. 10.5: *Naming ketones as acetyl and benzyl derivative.*

Problem: Write the name of following aldehydes/ketones.

a. $(CH_3)_2CHCH_2CH_2CH=O$ b. $CH_3CH_2CH=CHCH=O$

c.

d. H_3C $CHCH_2$ H_3C CH_3 O

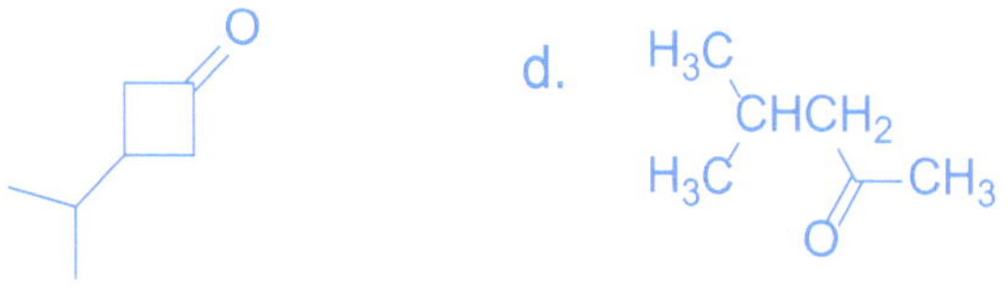

Solution: a. 4-methylpentanal, b. 2-pentenal (or pent-2-enal), c. 2-isopropylcyclobutanone, d. 4-methyl-2-pentanone (or 4-methlpentan-2-one).

10.3 Preparation of Aldehydes and Ketones

One of the best methods for preparation of aldehydes and ketones – the oxidation of alcohols, has been described previously (Chapter 8, 8.6e). Primary alcohols give aldehydes on oxidation with PCC in dichloromethane. Secondary alcohols are oxidized to ketones either with PCC in dichloromethane or with Jone's reagent.

PCC
CH_2Cl_2

PCC
CH_2Cl_2

Hydration of alkynes can also be used to prepare ketones.

H_2O, H^+, Hg^{2+}

A ketone with C=O attached to benzene ring such as acetophenone is prepared by the Friedel-Crafts acylation reaction of benzene (Chapter 5).

CH_3COCl-$AlCl_3$
80 °C

10.4 Carbonyl Group: Geometry and Chemistry

The carbon-oxygen double bond in aldehydes and ketones is similar in some respects to carbon-carbon double bond in alkenes. The carbon atom is *sp*2-hybridized (**Figure 10.6**). It forms three sigma bonds – one with oxygen and two with other atoms. The pi bond is formed by overlap of an unhybridized *p*-orbital of carbon with *p*-orbital of oxygen. The oxygen has two non-bonding pairs of electrons, which occupy its remaining two orbitals. The three atoms attached to carbonyl carbon lie in a plane with a bond angle of 120°. The C=O bond length is 1.24A°, obviously shorter than C-O bond length in alcohols and ethers.

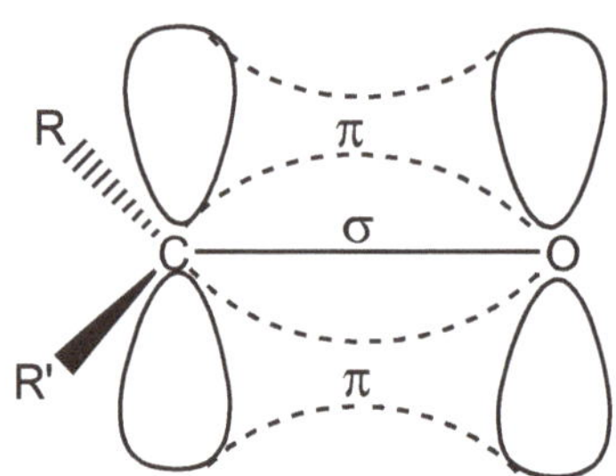

Fig. 10.6: *Formation of π bond in carbonyl carbon.*

The carbon-oxygen double bond is a highly polarized bond because of the high electronegativity difference between carbon and oxygen. Since the oxygen has high electronegativity, it carries a partial negative charge whereas the carbon carries a partial positive charge (**Figure 10.7**). Since carbon is positively charged, it is electrophilic and reacts with nucleophiles. We will see these reactions in Section 10.6 of this chapter.

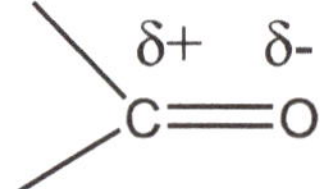

Fig. 10.7: *Polarization of C=O bond.*

10.5 Physical Properties

The physical properties are affected by polarization of the carbonyl group. They boil at higher temperature than hydrocarbon of about same molecular weight but at lower temperature than alcohol of about same molecular weight. In fact, hydrocarbons polarize temporarily by van der Waals attraction but the C=O has permanent polarization leading to dipole – dipole interaction between the molecules. The latter interaction is much stronger than van der Waals attraction but weaker than H-bond between alcohol molecules. As a result, aldehydes and ketones boil at higher temperature than hydrocarbon of same molecular weight but at lower temperature than alcohol of same molecular weight.

Carbonyl compounds with lower molecular weight are soluble in water because of the polar nature of the carbonyl group and because the molecules of these compounds can form H-bond with water.

10.6 Chemical Properties

a. Nucleophilic Addition on Carbon – Oxygen Double Bond

Nucleophilic addition reactions are the most common reactions of aldehydes and ketones. Nucleophiles react at the carbon atom of the carbon-oxygen double bond because the carbon atom carries a partial positive charge (**Figure 10.8**). After the attack of nucleophile, the π electrons of the C=O bond move to oxygen and sp^2-hybridized carbonyl carbon becomes sp^3-hybridized carbon. The negatively charged oxygen is neutralized by addition of a proton from the solvent.

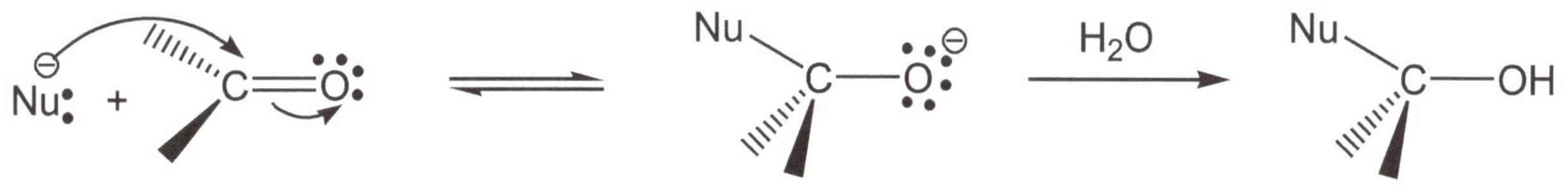

$$Nu^{\ominus} + >C=O \rightleftharpoons Nu-C-O^{\ominus} \xrightarrow{H_2O} Nu-C-OH$$

Fig. 10.8: *Reaction of nucleophile at carbonyl carbon.*

Carbonyl compounds are weak Lewis bases because of unshared pairs of electrons on oxygen atom. Acids, thus, catalyze the nucleophilic addition by protonating the oxygen atom of the carbonyl group (**Figure 10.9**).

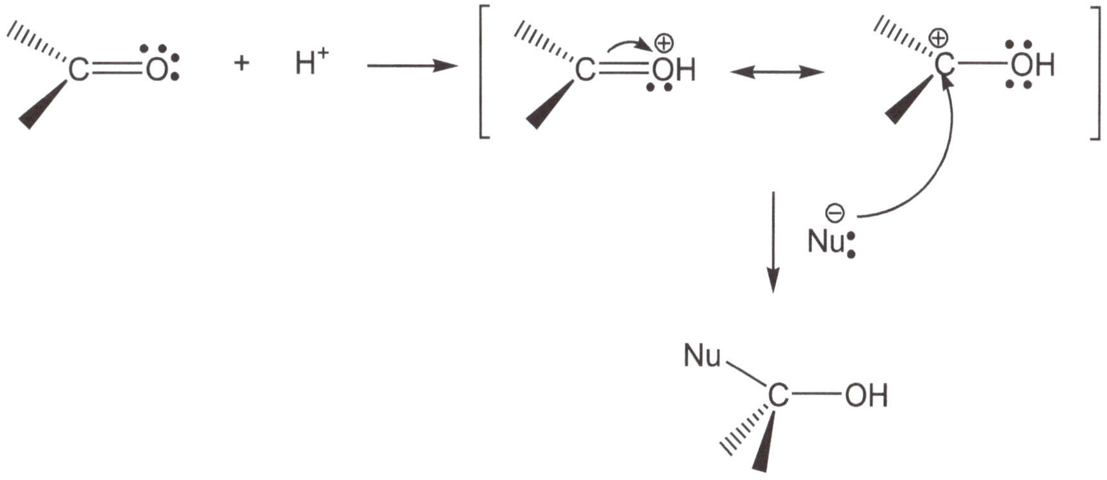

$$>C=O + H^+ \longrightarrow \left[>C=\overset{\oplus}{O}H \longleftrightarrow >\overset{\oplus}{C}-OH \right] \xrightarrow{Nu^{\ominus}} Nu-C-OH$$

Fig. 10.9: *Carbonyl compounds as Lewis bases.*

Aldehydes are usually more reactive in nucleophilic addition reactions than ketones due to steric and electronic factors. The carbonyl carbon is more crowded in ketones (two organic groups) than in aldehydes (one carbon group and one hydrogen atom). The alkyl groups are usually electron-donating in comparison to hydrogen atom and tend to neutralize the positive charge on carbonyl carbon, decreasing its reactivity towards nucleophiles.

Addition of water: Hydration

Water is an oxygen nucleophile and adds to aldehydes and ketones forming hydrate. It is a reversible reaction and hydrates lose water molecule to form aldehydes and ketone again. The equilibrium of the reaction favors the most stable product which is usually the carbonyl compounds. Trichloroacetaldehyde, commonly known as chloral produces stable hydrate which is used in medicines as sedative.

H_3C C=O H + H—OH ⇌ HO C—OH H_3C H

Addition of alcohols: Acetal formation

Alcohols like water are also weak oxygen nucleophiles and add to aldehydes and ketones in the presence of an acid as a catalyst to form acetals, compounds with two –OR group attached to same carbon atom (originally of C=O). The addition of first molecule of alcohol leads to the formation of a compound called hemiacetal which contains an ether and a hydroxyl group on same carbon. In the presence of excess alcohol, hemiacetal reacts further to form acetals.

O=C → (ROH, + H^+) → HO C OR → (ROH, + H^+) → RO C OR

Aldehyde or ketone — Hemiacetal — Acetal

The hemiacetal is formed in a similar manner as shown in Figure 10.9. The hydroxyl group of hemiacetals is protonated to form an intermediate **A**. The loss of water molecule from intermediate **A** generates a carbocation **B** that is attacked by alcohol forming another intermediate **C**. This intermediate loses a proton to form acetal.

Hemiacetal

Hemiacetal

A

Carbocation **B**

Acetal

C

Addition of hydrogen cyanide

Hydrogen cyanide adds to carbonyl group of aldehydes and ketones in the presence of a hydroxide base such as potassium hydroxide to form cyanohydrins, a class of compounds with a cyano (nitrile) group and a hydroxyl group attached to the same carbon. For example, the addition of hydrogen cyanide to acetone forms acetone cyanohydrin. The nucleophile is a carbon nucleophile, CN (cyanide) which attaches to the carbonyl carbon and the hydrogen attaches to the oxygen atom of the carbonyl group. The IUPAC name of the cyanohydrin of acetone is 2-hydroxy-2-methylpropanenitrile.

$$(CH_3)_2C{=}O + H{-}C{\equiv}N \xrightarrow{KOH} NC{-}C(OH)(CH_3){-}CH_3$$

Acetone — Cyanohydrin of acetone

Addition of Grignard reagents

The addition of a Grignard reagent to aldehydes and ketones is also an example of addition of a carbon nucleophile to carbonyl group. Addition to formaldehyde gives primary alcohols and to other aldehydes gives secondary alcohols. Ketones form tertiary alcohols by addition of the Grignard reagents. The organic group of the Grignard reagent reacts as nucleophile and attacks the carbonyl carbon forming an alkoxide ion. The intermediate alkoxide ion on protonation yields the alcohol.

$$R{-}MgX + H_2C{=}O \longrightarrow H_2C(R){-}OMgX \xrightarrow[H^+]{H_2O} H_2C(R){-}OH$$

formaldehyde — 1° alcohols

$$R{-}MgX + HR'C{=}O \longrightarrow HR'C(R){-}OMgX \xrightarrow[H^+]{H_2O} HR'C(R){-}OH$$

other aldehydes — 2° alcohols

$$R{-}MgX + R'_2C{=}O \longrightarrow R'_2C(R){-}OMgX \xrightarrow[H^+]{H_2O} R'_2C(R){-}OH$$

Ketones — 3° alcohols

Addition of acetylides

Acetylides react with carbonyl compounds in a manner like the Grignard reagents producing the corresponding alcohols.

$$R_2C{=}O + HC{\equiv}C^-Na^+ \text{ (sodium acetylide)} \longrightarrow HC{\equiv}C{-}CR_2{-}O^{\ominus} \xrightarrow[H_2O]{H^+} HC{\equiv}C{-}CR_2{-}OH$$

Addition of amines (nitrogen nucleophiles)

Amines have an unshared electron pair on its nitrogen atom. It, thus, behaves as a nitrogen nucleophile. The nucleophilic addition of amines to C=O bond of aldehydes and ketones gives an addition product. Elimination of a water molecule from the addition product gives **imines**.

$$R_2C{=}O + H_2\ddot{N}{-}R \rightleftharpoons [R_2C(OH){-}NH{-}R] \xrightarrow{-H_2O} R_2C{=}N{-}R$$

Imines

Problem:

Benzaldehyde (C_6H_5CHO) reacts as follows:

a ← CH_3MgBr / dry ether

b ← HCN

c ← $2\ CH_3OH / H^+$

d ← $H_2NHN{-}C_6H_5 / H^+$

Solution:

a $C_6H_5{-}CH(CH_3)(OH)$ b $C_6H_5{-}CH(CN)(OH)$ c $C_6H_5{-}CH(OCH_3)_2$ d $C_6H_5{-}CH{=}N{-}NH{-}C_6H_5$

b. Reduction of Aldehydes and Ketones

The reduction of aldehydes and ketones gives primary and secondary alcohols, respectively. The most common reagents used for reduction is lithium aluminum hydride. However, other metal hydrides are also known as reducing agents for aldehydes and ketones, e.g., sodium borohydride in combination with various transition metal salts. The polarization of the metal hydrogen bond takes place in a way that hydrogen atom gets negatively charged and the metal gets positively charged. The negatively charged hydrogen, now a hydride nucleophile, attacks the carbonyl carbon forming aluminum alkoxide which is subsequently hydrolyzed by a dilute acid to give the alcohol. For example, cyclohexanone gets reduced to cyclohexanol.

$$RR'C{=}O \xrightarrow[H^+,\ H_2O]{LiAlH_4} RR'CH{\cdot}OH$$

Cyclohexanone $\xrightarrow{H^{\ominus}\text{—}AlH_3}$ $[C_6H_{10}(H)\text{—}O^{\ominus}\text{—}AlH_3]\ Li^{\oplus}$ $\xrightarrow[H_2O]{H^+}$ Cyclohexanol

c. Oxidation of Aldehydes and Ketones

Aldehydes are easily oxidized to carboxylic acids with same number of carbon atoms. Many oxidizing agents such as Ag_2O, $KMnO_4$ and CrO_3 are used to accomplish the reaction. Ketones are very less reactive in oxidation and hence require special conditions for the reaction. Taking advantage of the difference in reactivity of aldehydes and ketones, aldehydes are distinguished from ketones in laboratory by the Tollens' silver mirror test. In this test, the silver-ammonia complex ion gets reduced by aldehydes to metallic silver called "*silver mirror*". Ketones fail to give this test.

$$H_3C\text{—}CH_2\text{—}CH_2\text{—}CHO \xrightarrow[\text{(Jones' reagent)}]{CrO_3,\ H^+} H_3C\text{—}CH_2\text{—}CH_2\text{—}COOH$$

$$R{-}\underset{H}{C}{=}O + 2\,Ag(NH_3)_2^+ + 3\,HO^- \longrightarrow R{-}\overset{O^\ominus}{C}{=}O + 2\,Ag^+\downarrow + 4\,NH_3^+\uparrow + 2\,H_2O$$

silver-ammnia complex ion (colorless) — silver mirror

10.7 Keto-Enol Tautomerism

Tautomerism (from the Greek word *tauto* meaning the same, and *meso* meaning the part) is a type of structural isomerism in which the two isomers differ only in the position of a hydrogen atom and a double bond and are called tautomers. A dynamic equilibrium exists between the two forms of which more stable form dominates the equilibrium.

Aldehydes and ketones having an α-hydrogen atom, a hydrogen attached to the carbon next to carbonyl carbon (-CH-C=O) are capable to exhibit tautomerism. The two tautomers are called *keto* form and *enol* form (ene + ol) (**Figure 10.10**). Most simple aldehydes and ketones, however, exist mainly in the keto form.

α-Hydrogen

Keto form ⇌ Enol form

Fig. 10.10: Keto-enol tautomerism in a ketone.

10.8 Acidity of α-Hydrogen atoms

The α-hydrogen in a carbonyl compound is more acidic in comparison to a hydrogen which is bonded to a carbon atom which has no adjacent carbonyl group. The higher acidity of the α-hydrogen is exploited to prepare different types of organic compounds. This hydrogen is abstracted by a base resulting into generation of a negatively charged intermediate called *enolate* (**Figure 10.11**) which can react with different electrophiles.

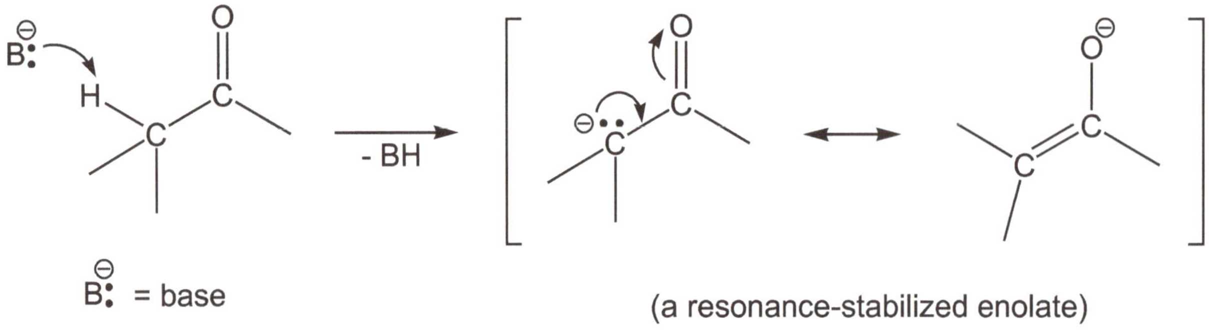

B: = base

(a resonance-stabilized enolate)

Fig. 10.11: *Generation of enolates.*

The reasons for the higher acidity of α-hydrogen are i) the bonding electrons are displaced toward the positively charged carbonyl carbon (**Figure 10.12**), and ii) the carbanion formed is resonance-stabilized (shown above).

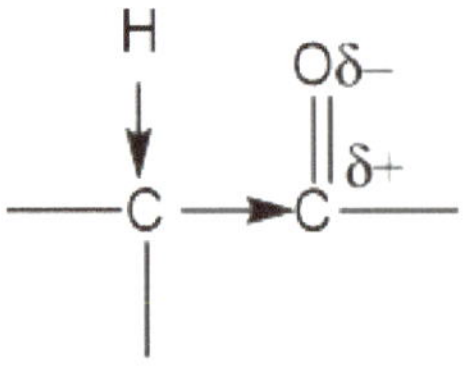

Fig. 10.12: *Displacement of bonding electrons towards carbonyl carbon.*

10.9 Aldol Reaction

The aldol reaction is the reaction of enolate ions with aldehydes or ketones to give β-hydroxycarbonyl compounds or α,β-unsaturated carbonyl compounds. The simplest example of an aldol reaction is the reaction between two molecules of acetaldehyde in the presence of a catalytic amount of aqueous sodium hydroxide to form a syrupy liquid called *aldol* (*ald* from aldehyde and *ol* from alcohol). The enolate ion, generated by abstraction of an α-hydrogen as a proton from one of the acetaldehyde molecules react with carbonyl group of another acetaldehyde molecule forming an alkoxide that gets neutralized by a proton from the solvent water to give the final product, a β-hydroxycarbonyl compound. Thus, it is basically an addition reaction and hence called aldol addition. The aldol reaction between the two molecules of the same compound is known as self-aldol addition. All the steps in mechanism are in equilibrium that lies towards far right in the final step if the reaction is of aldehydes.

$$CH_3CHO + CH_3CHO \xrightarrow{KOH} \underset{\text{Aldol}}{CH_3CH(OH)CH_2CHO}$$

$$CH_3CHO + HO^- \rightleftharpoons {}^-CH_2CHO + H_2O \quad \text{Step 1}$$

$$CH_3CHO + {}^-CH_2CHO \rightleftharpoons CH_3CH(O^-)CH_2CHO \quad \text{Step 2}$$

enolate nucleophile — An alkoxide

$$CH_3CH(O^-)CH_2CHO + H_2O \rightleftharpoons CH_3CH(OH)CH_2CHO + HO^- \quad \text{Step 3}$$

Aldol

The reaction may occur between two different aldehydes or ketones but one of the carbonyl compounds must have an α-hydrogen atom. The reaction between two different aldehydes or ketones is called mixed aldol addition. In the equation shown below, benzaldehyde has no α-hydrogen atom. The enolate, obtained from acetaldehyde adds onto the CH=O group of the benzaldehyde to form product **A**. The enolate, however, can also react with another molecule of acetaldehyde (self-aldol) to form product **B**. There are different ways to control the formation of product in aldol reactions that you will learn at higher levels.

$$C_6H_5CHO + CH_3CHO \xrightarrow{KOH} C_6H_5CH(OH)CH_2CHO + CH_3CH(OH)CH_2CHO$$

Benzaldehyde — Mixed-aldol product (**A**) — Self-aldol product (**B**)

The aldols readily dehydrate to form α,β-unsaturated carbonyl compounds in many cases. Aldol reaction, in fact, is a very important method for preparing α,β-unsaturated carbonyl compounds. Aldol reactions also take place in the presence of an acid. Under acidic conditions the reaction proceeds with protonation of carbonyl group.

$$C_6H_5CH(OH)CH_2CHO \xrightarrow[-H_2O]{} C_6H_5\overset{\beta}{C}H{=}\overset{\alpha}{C}H{-}CHO$$

Cinnamaldehyde
(an α,β-unsaturated aldehyde)

Problem: How many aldol addition products are theoretically possible on treating a mixture of acetaldehyde and propionaldehyde?

Strategy: In this case, two enolates will be formed. Each enolate can react with the aldehyde from which it is generated (self-aldol) and also with other aldehyde (cross-aldol).

Solution: Four aldol addition products can be formed in this case.

SUMMARY

1. Preparation of Aldehydes and Ketones

a. $R-CH_2-OH$ (1° alcohol) $\xrightarrow{PCC}$ $R(H)C=O$ (aldehyde)

b. $R-CH(R)-OH$ (2° alcohol) $\xrightarrow{PCC}$ $R_2C=O$ (ketone)

c. $R\text{-}C\equiv C\text{-}H \xrightarrow{H_2O,\ H^+,\ Hg^{2+}} R(H_3C)C=O$

d. $C_6H_6 \xrightarrow[AlCl3]{RCOCl} C_6H_5COR$

2. Nucleophilic Additions to Aldehydes and Ketones

$R_2C=O + ROH \underset{}{\overset{H^+}{\rightleftharpoons}} R-C(OH)(OR)-R$ (hemiacetal) $\underset{H^+}{\overset{ROH}{\rightleftharpoons}} R-C(OR)(OR)-R$ (acetal)

$R_2C=O + HCN \xrightarrow{NaOH\ (cat)} R-C(OH)(CN)-R$ (cyanohydrin)

$R_2C=O + R'\text{-}NH_2 \longrightarrow R_2C=N-R'$ (imine)

$$R_2C{=}O + R'\text{-}MgX \xrightarrow{\text{ether}} R{-}C(OMgX)(R'){-}R \xrightarrow{H_3O^+} R{-}C(OH)(R'){-}R$$

3. Mechanism of Nucleophilic Addition

$$Nu:^{\ominus} + R_2C{=}O \longrightarrow \left[Nu(R)_2C{-}O^{\ominus} \right] \xrightarrow{H_2O} Nu(R)_2C{-}OH$$

4. Oxidation to Carboxylic acids

$$R(H)C{=}O \xrightarrow{\text{Jone's reagent}} R(HO)C{=}O$$

5. Reduction to Alcohols

$$R_2C{=}O \xrightarrow[H_2\text{, catalyst, heat}]{LiAlH_4 \text{ or } NaBH_4} R{-}C(OH)(H){-}R$$

R = H, alkyl

6. Aldol reactions

$$2\ RCH_2CH{=}O \xrightarrow{\text{base}} RCH_2CH(OH)CH(R)CH{=}O$$

EXERCISES

1. Write the IUPAC name of the following compounds.

a. $CH_3CH(CH_3)CH_2CH{=}O$ b. 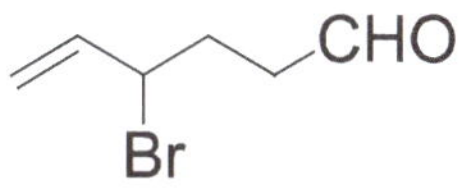 c. 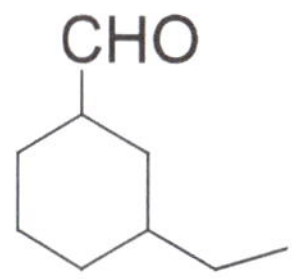 d.

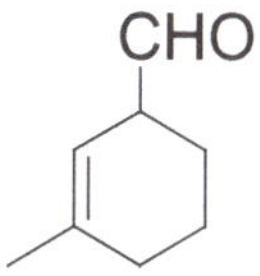

e. 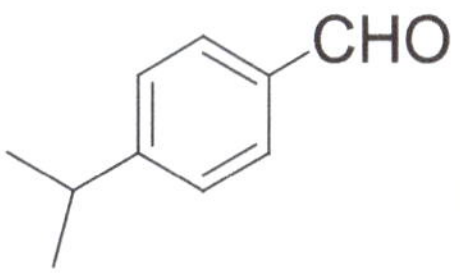 f. 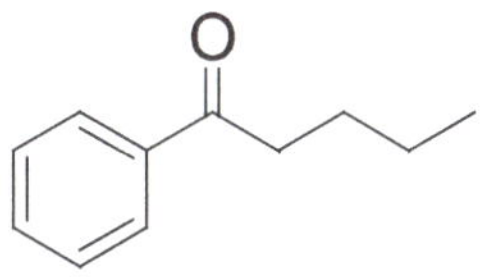 g. 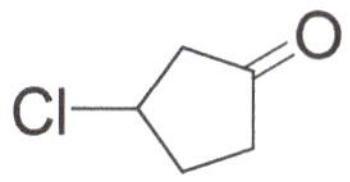 h.

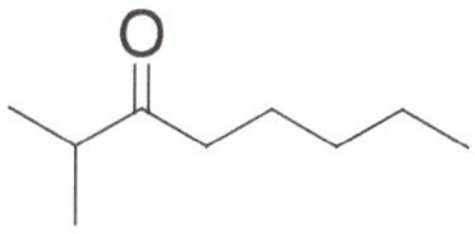

2. Write a structural formula for each of the following compounds containing either an aldehyde or a ketone functional group.

a. 3-methylpentanal b. 2- hexanone c. 2-bromobenzaldehyde

d. 3-chloropentenal e. 3-chlorocyclopent-1-enecarbaldehyde

f. 3-methyl-3-heptenal g. allyl methyl ketone h. 4-methylbenzphenone

i. 3-chlorocyclohex-2-enone j. benzyl ethyl ketone k. hept-3-ynal

3. Which of the compounds in Exercise 1 are chiral?

4. Write an equation for the synthesis of pentanal from an alcohol.

5. Write an equation for the preparation of following ketone by the Friedel-Crafts acylation.

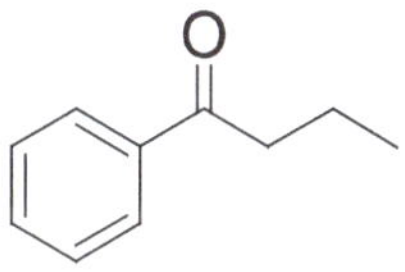

6. By writing the chemical equations, give two methods for the preparation of 2-heptanone.

7. Write the structure of the product that will be obtained by treatment of ketone in exercise 4 with sodium borohydride in ether as the solvent.

8. How can you distinguish between the following pairs of compounds by simple chemical test?

   a. 2-cyclopentenone and cyclopentanone

   b. heptanal and 2-heptanone

   c. benzyl alcohol and benzaldehyde

9. Briefly explain the fact that the boiling points of butanal (76 °C) and 1-butanol (118 °C) are quite different but their solubility in water is almost similar.

10. Explain why pentane-2,4-dione shows stronger acidic character in comparison to phenol (pK_a 9.0; less than phenol pK_a 10.0).

11. Write a brief note on each of the following:

    a. keto-enol tautomerism b. aldol condensation c. acetals and hemiacetals

12. Write a chemical equation for the reaction of *p*-tolualdehyde with each of the following reagent. Identify the class of the compound.

    a. methanol in the presence of a protic acid.

    b. sodium cyanide in the presence of a protic acid.

    c. isopropyl amine, $(CH_3)_2CHNH_2$

    d. phenyl hydrazine ($PhNHNH_2$)

    e. hydroxylamine (NH_2OH)

13. Provide the structure of aldehyde/ketone and other reagent required for the preparation of following compounds.

a. 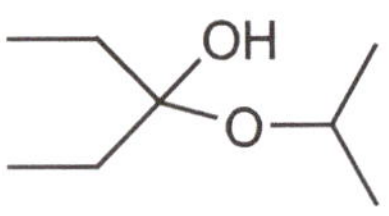 OH, O b. 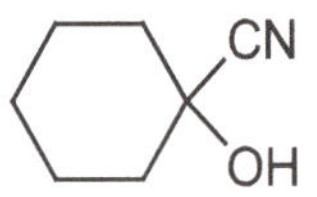 CN, OH c. 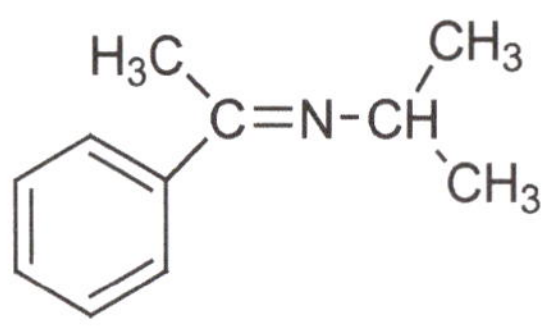 H_3C, C=N-CH, CH_3, CH_3 d. 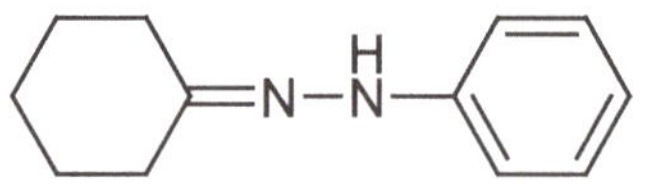 =N–N(H)–

14. Write the structure of the product when benzyl magnesium bromide ($PhCH_2MgBr$) is reacted with following aldehydes/ketones and the reaction mixture is subjected to acidic hydrolysis.

    a. formaldehyde

    b. 1-butanal

    c. 2-butanone

    d. benzaldehyde

    e. acetophenone

    f. 3-cyclohexenone

    g. cyclopentane carbaldehyde

15. Write the structure of Grignard reagents and aldehydes/ketones' combination that can be used for the synthesis of following alcohols.

    a. 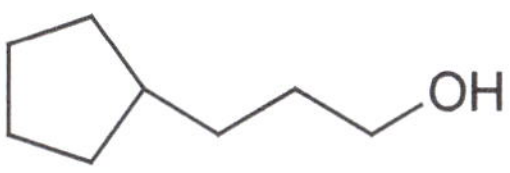 OH b. 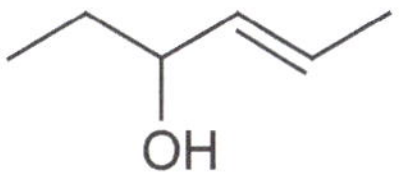 OH c. 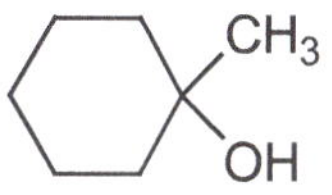 CH_3 OH

16. Using appropriate sodium acetylide and aldehyde/ketone write equations for the preparation of following alcohols.

    a. 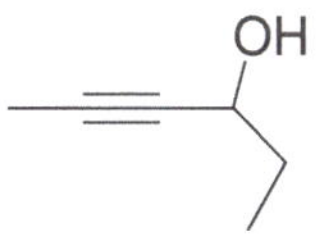 OH b. 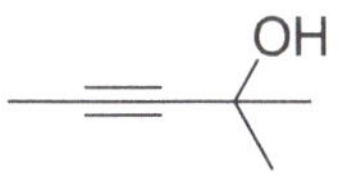 OH c. 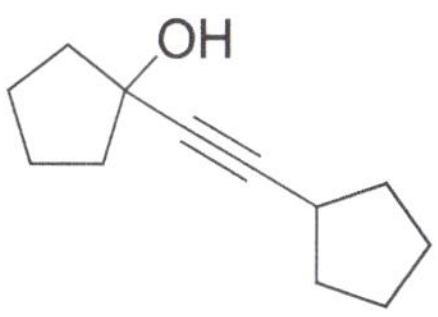 OH

17. Write the product of the following reactions.

    a. 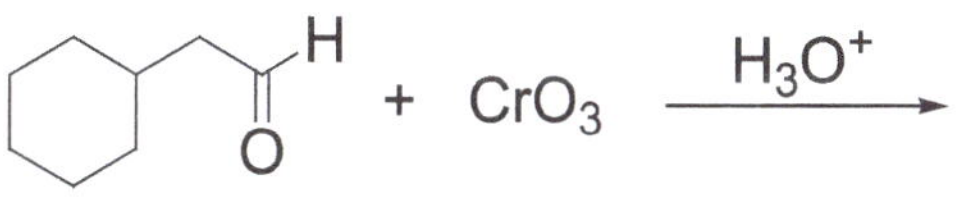 H O + CrO_3 $\xrightarrow{H_3O^+}$ b. 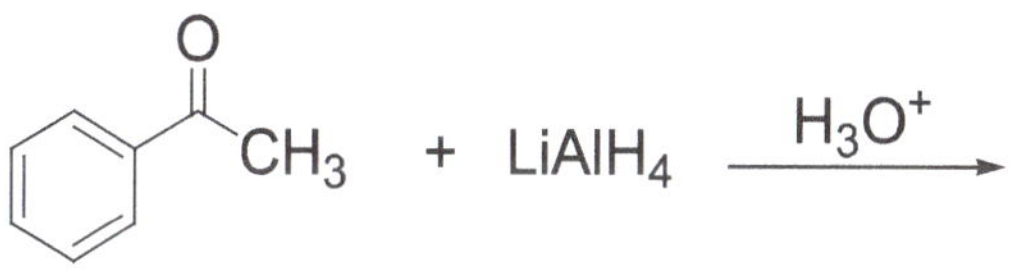 O CH_3 + $LiAlH_4$ $\xrightarrow{H_3O^+}$

    c. 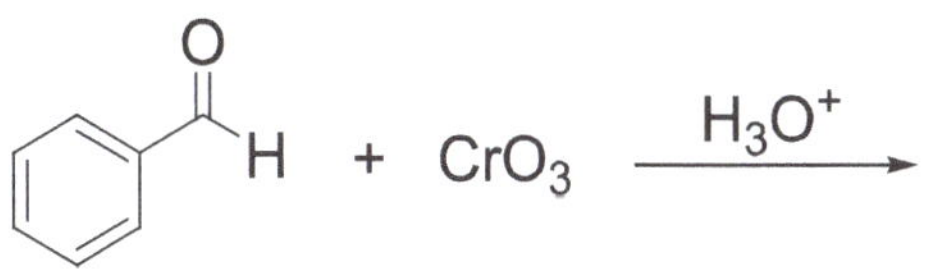 O H + CrO_3 $\xrightarrow{H_3O^+}$ d.  O H + $LiAlH_4$ $\xrightarrow{H_3O^+}$

18. Write the structure of the mixed aldol addition product that will be obtained on the reaction of *p*-tolualdehyde with propanal. Write the structure of the dehydration product as well.

19. Show stepwise mechanism of the formation of aldol in Exercise 18.

20. Complete the following reaction chart by providing the missing reagents and products.

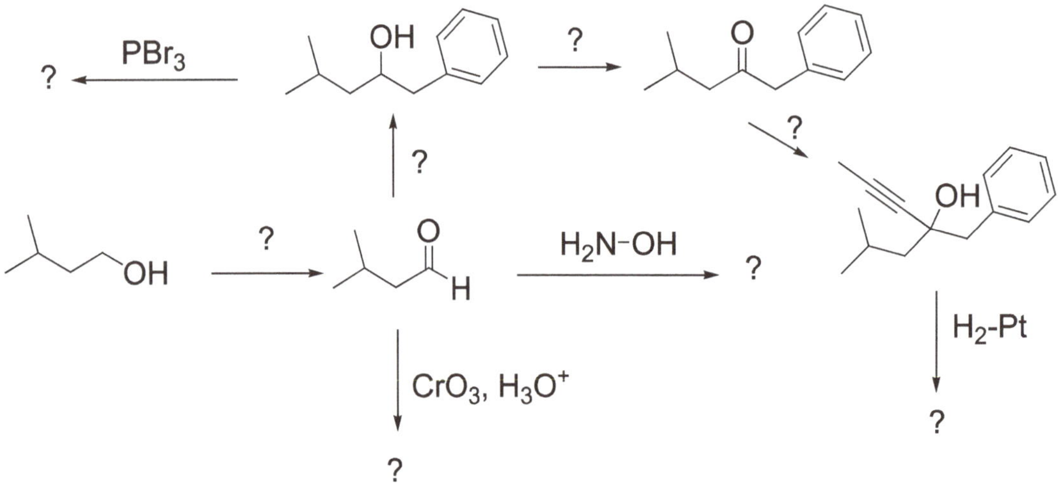

11 CARBOXYLIC ACIDS AND THEIR DERIVATIVES

Carboxylic acids are among the oldest class of compounds studied by organic chemists. They are abundant in natural resources. You have heard of acetic acid occurring in vinegar, tartaric acid occurring in tamarind, and citric acid occurring in lemons and oranges, and similar fruits that are called citrus fruits. They are characterized by the presence of carboxylic acid, -COOH, functional group (**Figure 11.1**). Thus, the general formula for this class is R-COOH where R can be alkyl or aromatic group. Several derivatives of organic compounds such as carboxylic esters, amides, acyl chlorides, and anhydrides are also important compounds.

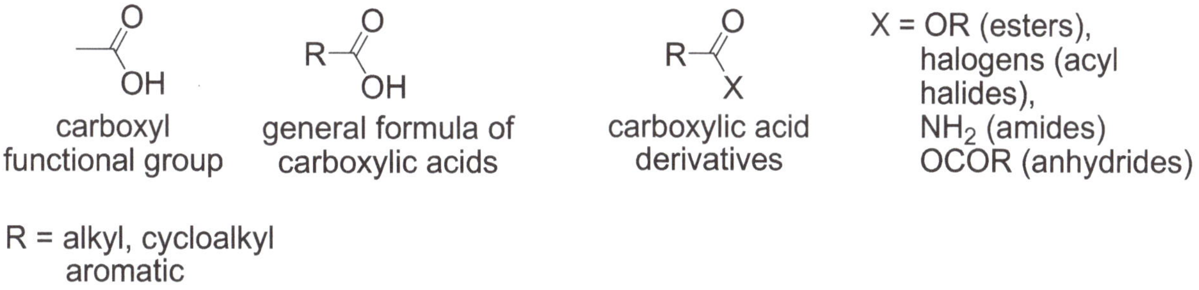

O OH
carboxyl functional group

R O OH
general formula of carboxylic acids

R O X
carboxylic acid derivatives

X = OR (esters),
halogens (acyl halides),
NH_2 (amides)
OCOR (anhydrides)

R = alkyl, cycloalkyl
aromatic

Fig. 11.1: *General formula of carboxylic acids and its derivatives.*

11.1 Nomenclature of Carboxylic acids

Many carboxylic acids have common names such as acetic acid, cinnamic acid, salicylic acid, phthalic acid, valeric acid, lactic acid, tartaric acid, etc. Many of these names are derived from the Greek or Latin words that indicate their occurrence. According to the IUPAC system, the carboxylic acids are named by replacing the "e" at the end of the parent carbon chain with "oic acid". For example, CH_3COOH is names as ethanoic acid (recall: two-carbon chain alkane is ethane). In cyclic carboxylic acids, the word "carboxylic acid" is put as suffix after the parent ring. The substituents are given minimum possible number. Some examples are shown below (**Figure 11.2**).

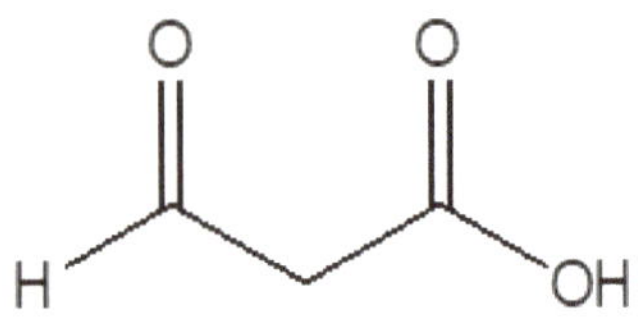

3-oxopropanoic acid

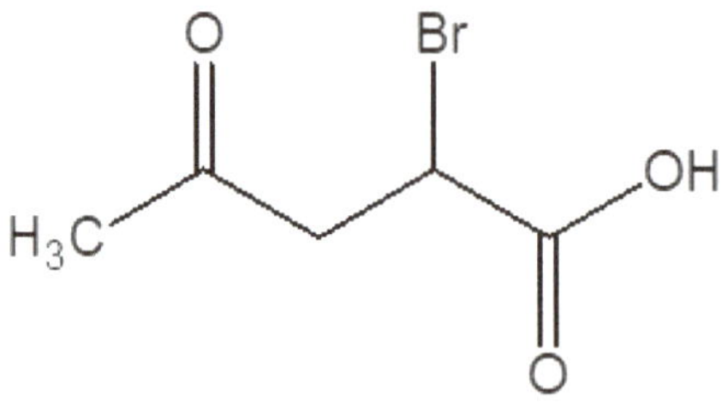

2-bromo-4-oxopentanoic acid

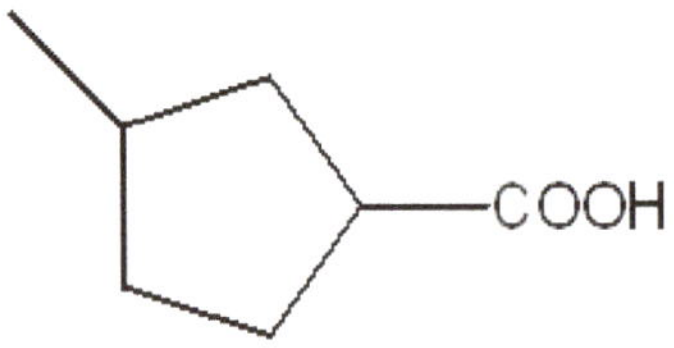

3-methylcyclopentanecarboxylic acid

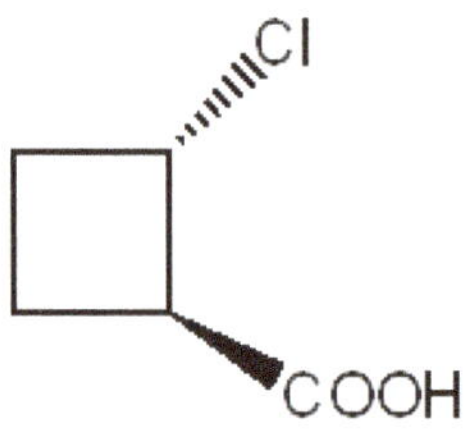

trans-2-chlorocyclobutanecarboxylic acid

Fig. 11.2: *Naming carboxylic acids.*

Problem: Name the following carboxylic acids which contain some other functional group or branch as well.

a. 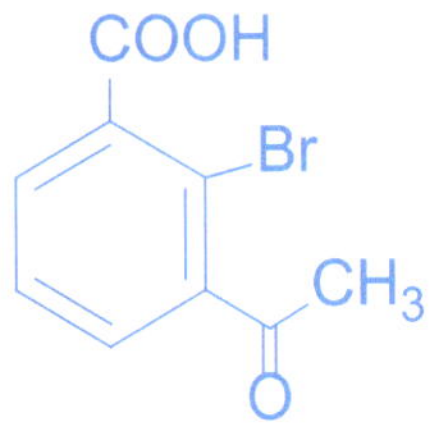 b. 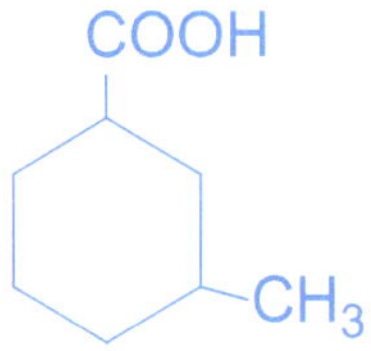 c.

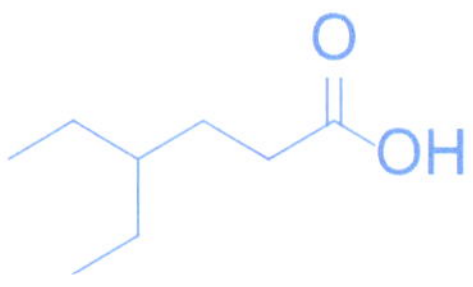

Solution: a. 3-acetyl-2-bromobenzoic acid, b. 3-methylcyclohexanecarboxylic acid, c. 4-ethylhexanoic acid

11.2 Preparation of Carboxylic acids

a. Oxidation of Alcohols and Aldehydes

Oxidation of primary alcohols using Jone's reagent (chromium trioxide in aqueous sulfuric acid and acetone) forms aldehydes. Aldehydes get further oxidized to carboxylic acids.

$$R-CH_2-OH \xrightarrow[\text{acetone}]{CrO_3,\ H^+} \left[R-\underset{H}{C}=O\right] \xrightarrow[\text{acetone}]{CrO_3,\ H^+} R-\underset{OH}{C}=O$$

b. Oxidation of Alkyl Side Chains on Aromatic Rings

The methyl, ethyl or isopropyl groups on aromatic rings can be oxidized by potassium permanganate offering an easy method for the preparation of aromatic acids. For example, toluene is oxidized by potassium permanganate to get benzoic acid.

$$C_6H_5-CH_3 \xrightarrow{KMnO_4} C_6H_5-COOH$$

c. Reaction of Grignard Reagents with Carbon dioxide

Carboxylic acids can also be prepared from the Grignard reagents. The reaction of Grignard reagents with carbon dioxide leads to the formation of carboxylic acids. For example, the reaction of ethyl magnesium bromide with carbon dioxide gives propanoic acid.

$$C_2H_5MgBr + CO_2 \xrightarrow{\text{ether}} \left[O{=}\underset{C_2H_5}{C}{-}O^-MgBr^+\right] \xrightarrow{H_2O} \underset{\text{propanoic acid}}{O{=}\underset{C_2H_5}{C}{-}OH}$$

Problem: How can you prepare cyclohexane carboxylic acid using an appropriate Grignard reagent?

Strategy: Keep in mind that the functional group HO-C=O is coming from carbon dioxide. So, you must prepare the Grignard reagent from an appropriate halide that will be the one containing the carbon chain or ring present in the carboxylic acid.

Solution: For preparation of cyclohexane carboxylic acid, the Grignard reagent will be prepared from bromocyclohexane and reacted with carbon dioxide.

$$C_6H_{11}Br + Mg \xrightarrow{\text{ether}} C_6H_{11}MgBr \xrightarrow[\text{2. } H_2O]{\text{1. } CO_2} C_6H_{11}COOH$$

cyclohexane carboxylic acid

d. Hydrolysis of Organic Cyanides

Organic cyanides (nitriles) give carboxylic acids on acidic hydrolysis.

$$R—C\equiv N \quad + \quad 2\,H_2O \xrightarrow{\text{aq. HCl}} R—COOH \quad + \quad NH_4Cl$$

11.3 Physical Properties of Carboxylic acids

The first few members of the carboxylic acid groups are colorless liquids with sharp or unpleasant odors. Acetic acid, which constitutes about 5% of vinegar, gives vinegar its characteristic odor and flavor. Benzoic acid is a white crystalline solid.

Carboxylic acids are polar compounds like alcohols. They form hydrogen bonds with themselves (**Figure 11.3**) or with other molecules. Hydrogen bonding also explains the solubility of low molecular weight carboxylic acids in water.

Fig. 11.3: *Intramolecular hydrogen-bonding in carboxylic acids.*

Carboxylic acids ionize in water forming carboxylate ions and hydronium ions. Carboxylate ions are conjugate bases of carboxylic acids.

$$R—COOH \quad + \quad H_2O \rightleftharpoons R—COO^{\ominus} \quad + \quad H_3O^{\oplus}$$

The acidity constant, K_a is given by the following expression:

$$K_a = \frac{[RCOO^{\ominus}][H_3O^{\oplus}]}{[RCOOH]}$$

Both alcohols and carboxylic acids ionize by losing a proton. However, the carboxylic acids are much stronger acids than alcohols (see the K_a of ethanol and ethanoic acid below). The stronger acidic property of carboxylic acids is due to stabilization of carboxylate ions by resonance (**Figure 11.4**).

$$CH_3CH_2OH + H_2O \rightleftharpoons CH_3CH_2O^{\ominus} + H_3O^{\oplus} \quad K_a = 1.0 \times 10^{-16}$$

$$CH_3COOH + H_2O \rightleftharpoons CH_3COO^{\ominus} + H_3O^{\oplus} \quad K_a = 1.8 \times 10^{-5}$$

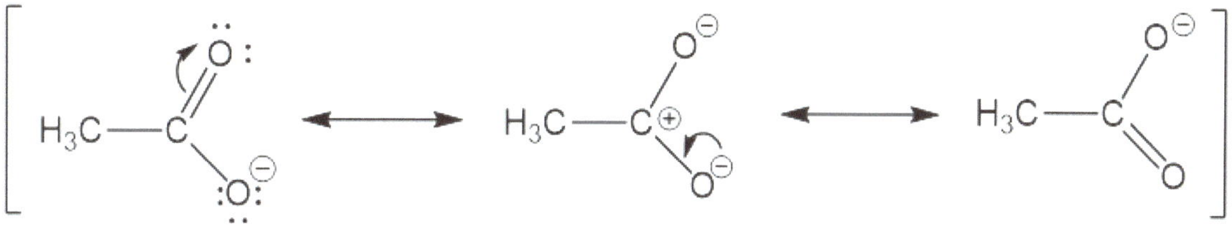

Fig. 11.4: *Resonance-stabilization of carboxylate ion.*

Among different carboxylic acids, the acidity depends on the groups attached to the carboxylic functional group. For example, chloroacetic acid is a stronger acid than acetic acid. Dichloroacetic acid is stronger acid than chloroacetic acid and trichloroacetic acid is even stronger than dichloroacetic acid. Increasing the chlorine atom on methyl group, thus, increases the acidity. Not only chlorine but all electron-withdrawing groups increase the acidity and electron-donating groups decrease the acidity. This can be explained by analyzing the inductive effect of chlorine on the corresponding carboxylate ions (**Figure 11.5**). The chlorine being more electronegative than carbon polarizes the C-Cl bond. It results in pulling electrons away from the carboxylate end towards chlorine. The more the number of negative charges more stable will be the carboxylate and hence stronger will be the acid.

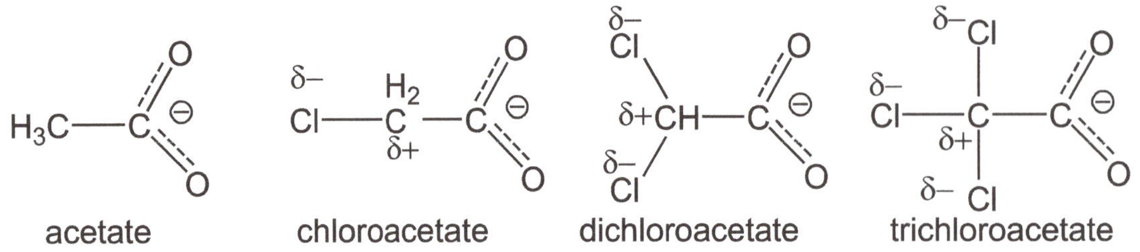

acetate chloroacetate dichloroacetate trichloroacetate

Fig. 11.5: *Polarization of C-Cl bond in chlorine-substituted acetates.*

11.4 Chemical Properties: Carboxylic Acid Derivatives

The reactions of carboxylic acids lead to many useful products. Acids react with strong bases to form salts. Carboxylic acids give alcohols on reduction. Carboxylic acids react with thionyl chloride or PCl_5, alcohols, and amines forming acid chlorides, esters, and amide derivatives, respectively, that will be dealt with separately because of their importance in organic synthesis.

a. Conversion of Acids to Salts

Treatment of carboxylic acids with strong bases such as sodium hydroxide or potassium hydroxide forms salts that can be isolated by evaporating the water. Salts of certain carboxylic acids are useful as *soaps* and *detergents*. In naming the salts, the cation is named first followed by the name of the carboxylate which is derived by replacing "ic" ending of the acid with "ate". Look at the structures of sodium alkanoates and potassium benzoate. Now a day we draw structures of carboxylic acid salts even without showing structures just like inorganic salts.

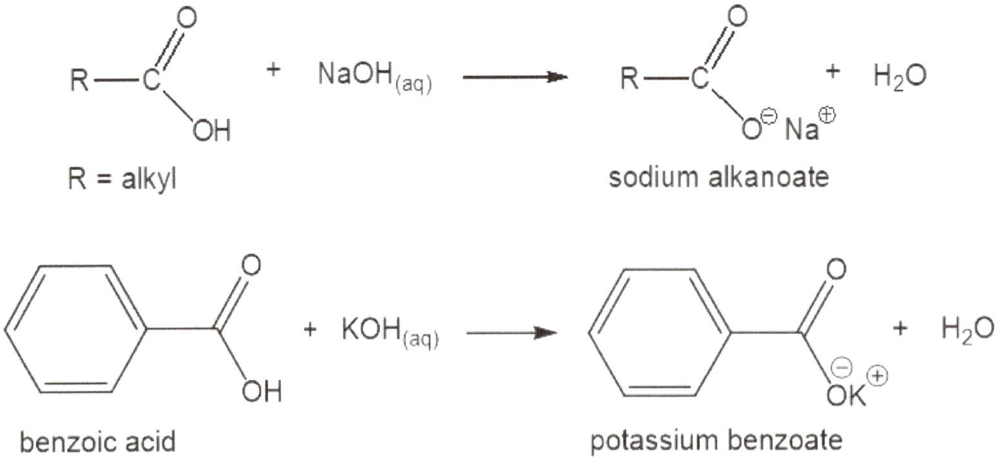

$$R{-}COOH + NaOH_{(aq)} \longrightarrow R{-}COO^{\ominus}Na^{\oplus} + H_2O$$

R = alkyl; sodium alkanoate

$$C_6H_5COOH + KOH_{(aq)} \longrightarrow C_6H_5COO^{\ominus}K^{\oplus} + H_2O$$

benzoic acid; potassium benzoate

b. Reduction of Carboxylic acids

You are familiar with oxidation of aldehydes to carboxylic acids. In reverse of this reaction, carboxylic acids can be transformed to alcohols. In the following example, the carboxylic group is selectively reduced by lithium aluminum hydride. This reagent does not reduce the C=C of the compound.

$$H_3C(H_2C)_7HC{=}CH(CH_2)_7COOH \xrightarrow[2.\ H_2O]{1.\ LiAlH_4} H_3C(H_2C)_7HC{=}CH(CH_2)_7CH_2OH$$

11.5 Acyl Halides

Acyl halides are the most reactive carboxylic acid derivatives. Acyl chlorides are more common and less expensive than bromides or iodides.

a. Preparations

These are prepared by the reaction of carboxylic acids with thionyl chloride or phosphorus pentachloride.

$$R{-}C({=}O){-}OH + SOCl_2 \longrightarrow R{-}C({=}O){-}Cl + HCl + SO_2$$

$$R{-}C({=}O){-}OH + PCl_5 \longrightarrow R{-}C({=}O){-}Cl + HCl + POCl_3$$

b. Reactions

Acyl halides react rapidly with most nucleophiles. Like other carboxylic acid derivatives, they can be hydrolyzed to the carboxylic acids. The reactions of acyl halides with alcohol and ammonia give carboxylic esters and primary amides, respectively.

$$R{-}C(=O)Cl + H_2O \longrightarrow R{-}C(=O)OH + HCl$$

$$R{-}C(=O)Cl + R^1{-}OH \longrightarrow R{-}C(=O)OR^1 + HCl$$

$$R{-}C(=O)Cl + 2\,NH_3 \longrightarrow R{-}C(=O)NH_2 + NH_4Cl$$

11.6 Carboxylic Esters

Esters are derived from carboxylic acids by replacing the hydroxyl group with an –OR (alkoxy, aryloxy) group (**Figure 11.6**). Esters are constituents in many fruits. In fact, the pleasant odor of many fruits is of the ester present in that fruit.

$$R{-}C(=O)OR$$

Fig. 11.6: *General structure of esters.*

a. Nomenclature of Carboxylic Esters

Start naming ester from the group directly attached to oxygen (alkyl, cycloalkyl, aryl) of the functional group [C(=O)O] (**Figure 11.7**). Then, number the chain starting from C=O of the functional group to the side opposite to which alkyl/cycloalkyl/aryl group is attached to oxygen atom. Delete "ic" from the end of the carboxylic acid name from which ester is derived and add "ate". If the chain has a multiple bond or any other group on it, include its name and position.

In the first example shown below, the group directly attached to the oxygen of the functional group ester is phenyl. So, phenyl comes first in the name. The other side is C(=O) attached to benzene ring. So, "ic" from benzoic (parent benzoic acid) is replaced with "ate" to get benzoate. Thus, the complete name of the ester is phenyl benzoate.

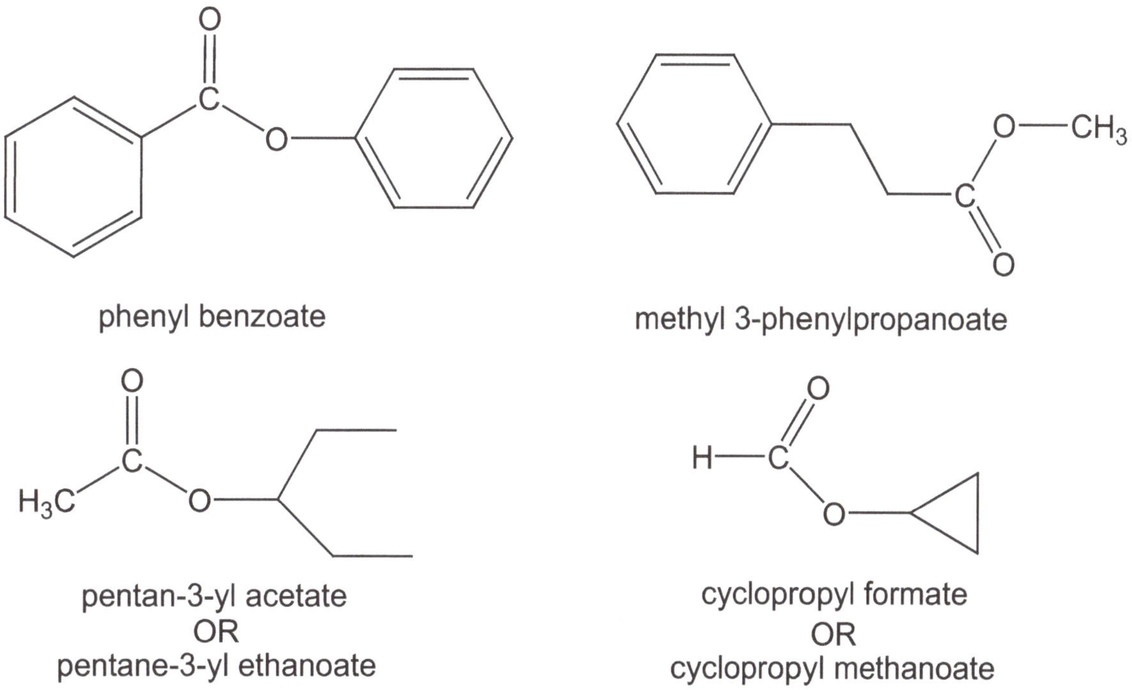

phenyl benzoate

methyl 3-phenylpropanoate

pentan-3-yl acetate
OR
pentane-3-yl ethanoate

cyclopropyl formate
OR
cyclopropyl methanoate

Fig. 11.7: *Naming carboxylic esters.*

b. Preparation of Carboxylic Esters

Esters are prepared by reaction of carboxylic acids with alcohols. The reaction is carried out in the presence of a protic acid as a catalyst.

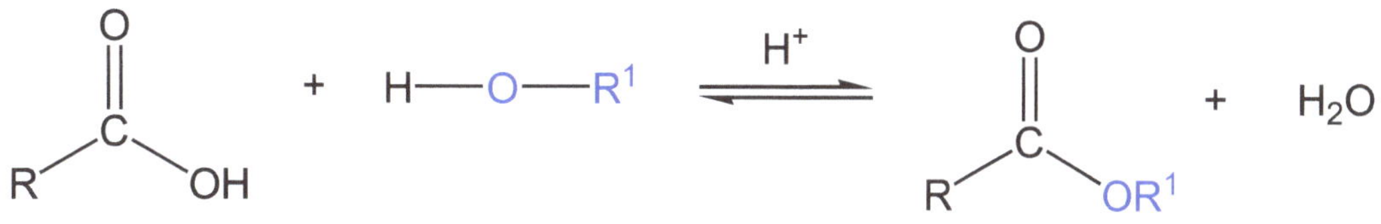

$$R-COOH + H-O-R^1 \overset{H^+}{\rightleftharpoons} R-COOR^1 + H_2O$$

c. Mechanism of Ester Formation

The formation of ester can be explained by the following six steps. The lone pairs on oxygen are shown only when necessary to explain the mechanism.

Step 1 protonation of the carbonyl oxygen in a reversible manner.

Step 2 the nucleophilic attack of alcohol on carbon atom of the protonated carboxylic group. This is the crucial step because a new C-O bond (ester bond) is formed in this step.

Step 3 a proton is lost from positively charged OH group in reversible manner to form a 1,1-diol.

Step 4 protonation of a hydroxyl group in 1,1-diol.

Step 5 removal of water (one product of the reaction) from the intermediate of step 4.

Step 6 deprotonation leading to formation of ester and regeneration of the catalyst.

d. Reactions of Carboxylic Esters

Like carboxylic acids, most reactions of esters are those which occur due to nucleophilic attack on carbonyl carbon atom forming a tetrahedral intermediate followed by loss of a leaving group. The rates of both the steps are enhanced by increasing the electron-withdrawing nature of the leaving group. A leaving group with electron-withdrawing property makes the carbonyl carbon more positive and is also a good leaving group.

Esters, in general, are less reactive than aldehydes and ketones towards nucleophiles. This is because of delocalization of the positive charge on the carbonyl carbon (**Figure 11.8**). The positive charge on carbon gets delocalized to the oxygen atom. As a result, the ester is more stable and less prone to attack by nucleophiles.

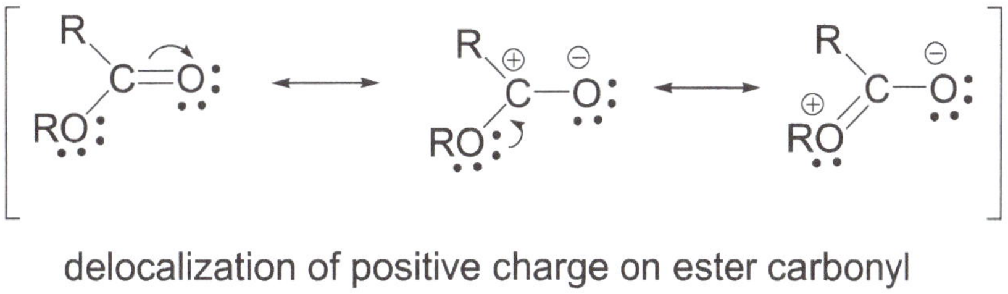

delocalization of positive charge on ester carbonyl

Fig. 11.8: *Delocalization of positive charge on ester carbonyl.*

Saponification of carboxylic esters

The hydrolysis of esters with a base is called saponification. This type of reaction is used to make soap from fats. The nucleophilic attack by hydroxyl group on the carbonyl carbon followed by removal of alkoxide produces carboxylic acid reversibly. In the next step, alkoxide ions abstract the proton from carboxylic acid to form alcohol and carboxylate ions.

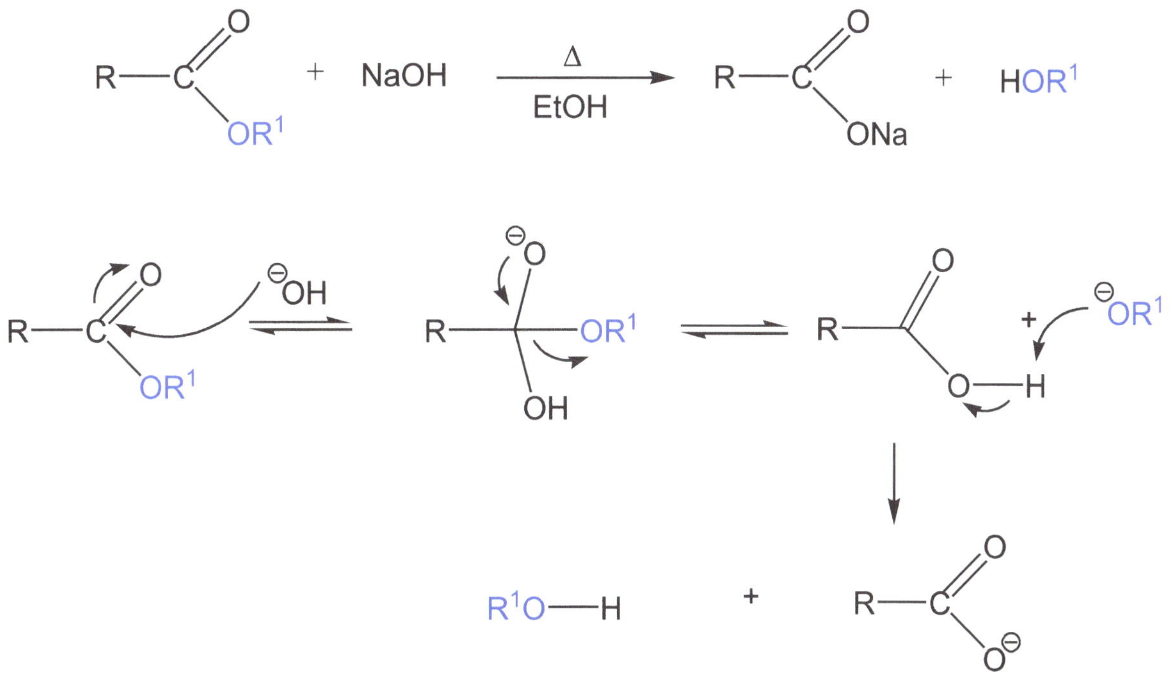

$$R-C(=O)OR^1 + NaOH \xrightarrow[EtOH]{\Delta} R-C(=O)ONa + HOR^1$$

$$R-C(=O)OR^1 + {}^{\ominus}OH \rightleftharpoons R-C(O^{\ominus})(OH)-OR^1 \rightleftharpoons R-C(=O)O-H + {}^{\ominus}OR^1 \rightarrow R^1O-H + R-C(=O)O^{\ominus}$$

Ammonolysis of carboxylic esters

Esters react with ammonia forming primary amides.

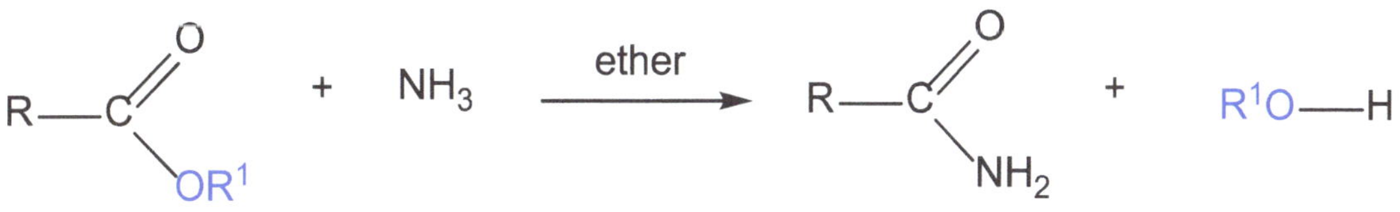

$$R-C(=O)OR^1 + NH_3 \xrightarrow{ether} R-C(=O)NH_2 + R^1O-H$$

Reactions of carboxylic esters with Grignard reagents

One mole of ester reacts with two moles of the Grignard reagents forming a *tertiary* alcohol. The reaction of the first mole of the Grignard reagent with ester forms a ketone which in turn reacts with another mole of the Grignard reagent affording the *tertiary* alcohol.

$$R-C(=O)OR^1 + 2\ \overset{\delta-}{R^2}-\overset{\delta+}{MgBr} \xrightarrow{\text{ether}} R\,C(OMgBr)(R^2)_2 \xrightarrow[H^{\oplus}]{H_2O} R\,C(OH)(R^2)_2 \ (\textit{tert}\text{-alcohol})$$

$$R-C(=O)OR^1 \xrightarrow{R^2MgBr} R\,C(OMgBr)(R^2)(OR^1) \xrightarrow{-\,R^1OMgBr} R-C(=O)-R^2 \ (\text{ketone}) \xrightarrow{R^2MgBr} R\,C(OMgBr)(R^2)_2$$

Reduction of carboxylic esters

An ester is reduced to alcohol using metal hydrides, e. g., sodium borohydride or lithium aluminum hydride.

$$R-C(=O)OR^1 \xrightarrow[\text{ether}]{LiAlH_4} R\text{-}CH_2OH + HOR^1$$

The α-hydrogen of esters: the Claisen condensation

You have studied about the acidity of α-hydrogen in aldehydes and ketones (section 10.8). The α-hydrogen in ester is also weakly acidic because it is adjacent to carbonyl group. The α-hydrogen can be removed by a strong base such as sodium hydride or sodium alkoxides to form an anion called ester enolate that is resonance-stabilized.

α-hydrogen

strong base

resonance-stabilized ester enolate

The ester enolate reacts as a carbon nucleophile. It reacts with carbonyl group of another ester molecule and expels alkoxide group to form β-keto ester in a reaction called *Claisen Condensation*.

$$H_3C\text{-}\overset{O}{\overset{\|}{C}}\text{-}OCH_2CH_3 + H\text{-}CH_2\text{-}\overset{O}{\overset{\|}{C}}\text{-}OCH_2CH_3 \xrightarrow[\text{2. } H_3O^+]{\text{1. } CH_3CH_2ONa \text{ in ethanol}} H_3C\text{-}\overset{O}{\overset{\|}{C}}\text{-}CH_2\text{-}\overset{O}{\overset{\|}{C}}\text{-}OCH_2CH_3 + CH_3CH_2OH$$

ethyl acetate + ethyl acetate → (ethyl acetoacetate a β-keto ester)

Problem: Show the mechanism of formation of the products of the Claisen condensation (Hint: First, generate the enolate from one molecule of ethyl acetate then show its nucleophilic reaction with carbonyl group of another molecule of ethyl acetate).

11.7 Amides

The compounds with general molecular formula R-$CONH_2$ are called primary amides (**Figure 11.9**). Replacement of one or both hydrogen atoms on nitrogen by other group leads to the formation of *secondary or tertiary amides*. Amides occur widely in nature. The most important amides are proteins. Amides are named by replacing 'oic acid' from the name of the corresponding acid with the 'amide'. Some examples are shown below.

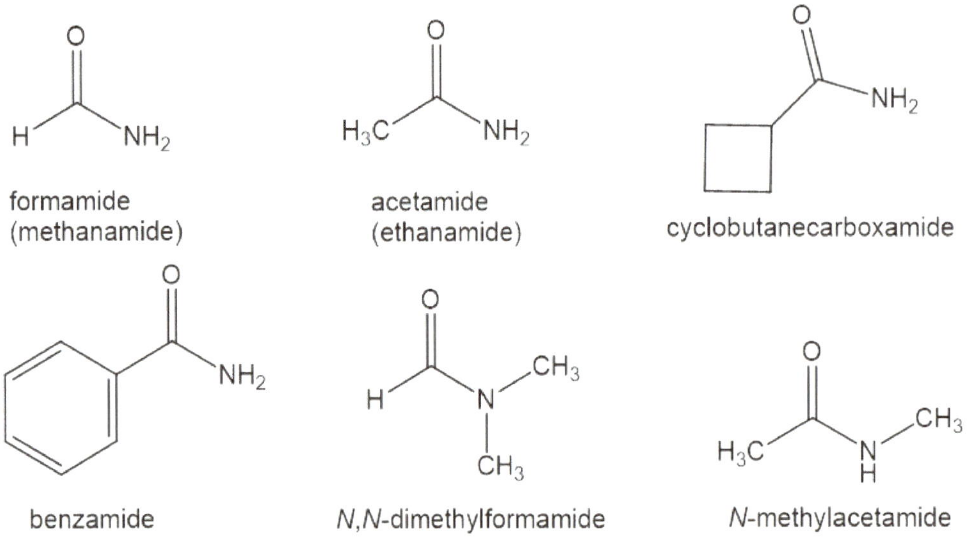

formamide (methanamide)
acetamide (ethanamide)
cyclobutanecarboxamide
benzamide
N,*N*-dimethylformamide
N-methylacetamide

Fig. 11.9: *Naming amides.*

The preparation of primary amides has been shown earlier by reactions of carboxylic esters and of acyl chlorides with ammonia. Similar reactions of carboxylic esters and of acyl chlorides with primary and secondary amines lead to the formation of secondary amides and tertiary amides, respectively.

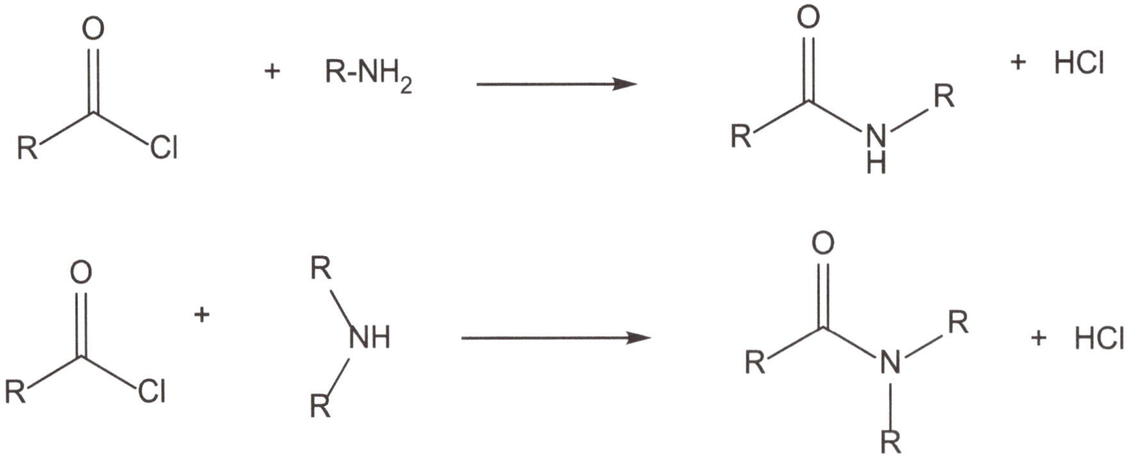

$$RCOCl + R\text{-}NH_2 \longrightarrow RCONHR + HCl$$

$$RCOCl + R_2NH \longrightarrow RCONR_2 + HCl$$

Amides are compounds having planar geometry. The C-N bond length in amides is only 1.32 A° that is much shorter than the normal carbon-nitrogen single bond length which is about 1.47 A°. The reason for this is attributed to the contribution from dipolar form due to resonance (**Figure 11.10**). In this form, the nitrogen is bonded to carbon by a double bond. This contribution is so strong that it restricts the rotation around the C-N bond. Also, due to contribution from dipolar structure the amides are highly polar compounds and form strong hydrogen bonds. Amides have exceptionally high boiling points. Alkyl substituents on nitrogen decrease the boiling point by decreasing the opportunity for hydrogen bonding.

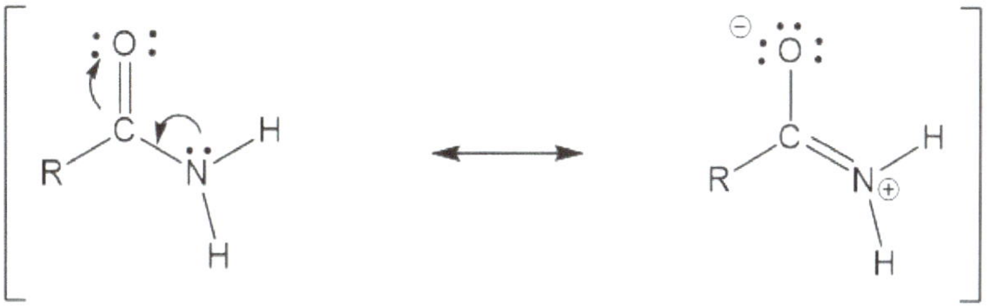

Fig. 11.10: *Dipolar structure of amide.*

Amides can be hydrolyzed to the corresponding acid. The reaction is slow and often requires prolonged heating or catalysis by acid or base. Reduction of amides with lithium aluminum hydride gives amines.

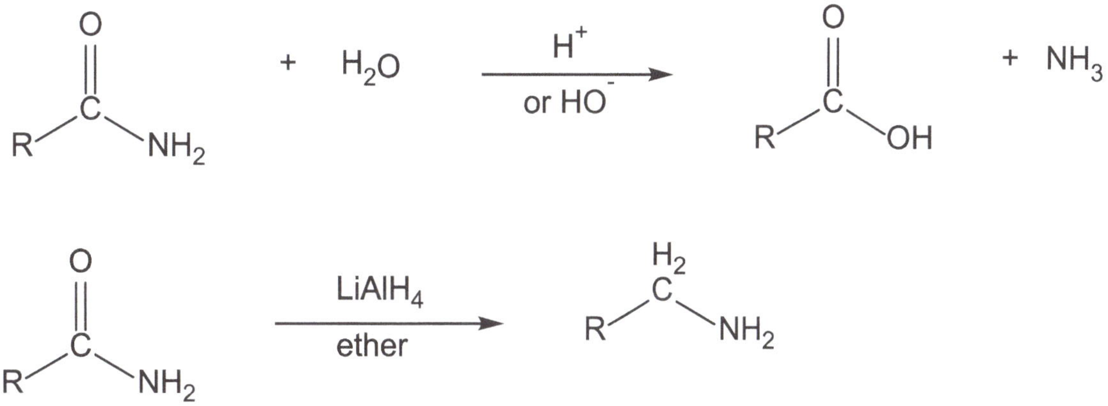

$$RCONH_2 + H_2O \xrightarrow[\text{or } HO^-]{H^+} RCOOH + NH_3$$

$$RCONH_2 \xrightarrow[\text{ether}]{LiAlH_4} RCH_2NH_2$$

11.8 Cyclic Esters and Amides

Cyclic esters are known as lactones. Most common lactones are five- or six-membered ring compounds although smaller or larger rings are possible. A macrocyclic lactone, erythromycin is widely used as an *antibiotic*, a class of drugs used to treat the bacterial infections. If an open-chain compound contains both a hydroxyl functional group and a carboxylic acid group, it can be transformed to a lactone by the reaction of OH group with COOH group. For example,

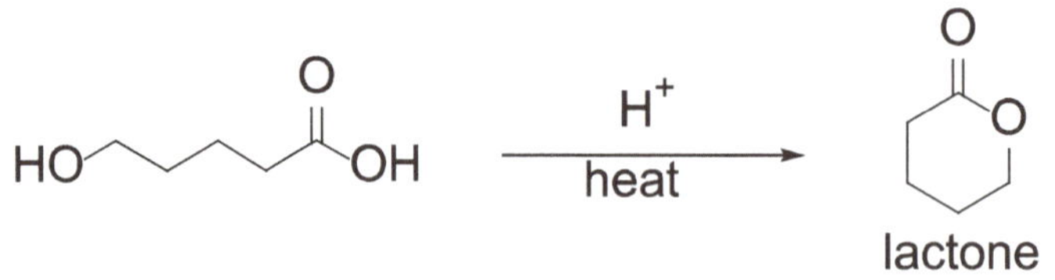

HO–(CH$_2$)$_4$–COOH $\xrightarrow[\text{heat}]{H^+}$ lactone

Cyclic amides are also biologically important compounds. Specially, the four-membered cyclic amides commonly known as β-lactams are well-known antibiotics (**Figure 11.11**). The famous antibiotics like penicillin and cephalosporin, that were isolated from microorganisms, contain β-lactam ring. The antibacterial activity of these drugs is attributed to the presence of β-lactam ring in them. Since the four-membered rings are highly strained and hence unstable, the synthesis and chemistry of these compounds are very interesting, and a great deal of research work has been done on such compounds.

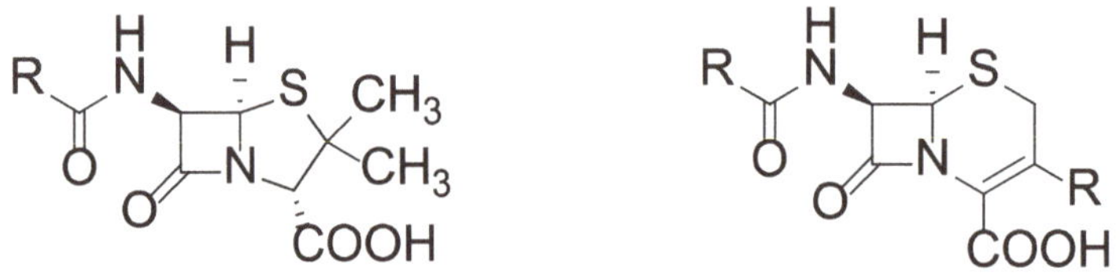

Fig. 11.11: *General structures of penicillin and cephalosporin.*

11.9 Acid Anhydrides

Acid anhydrides are derived by combining two same or different acids with elimination of a water molecule. A good method for preparation of anhydrides involves the reaction of an acyl chloride with a carboxylate salt.

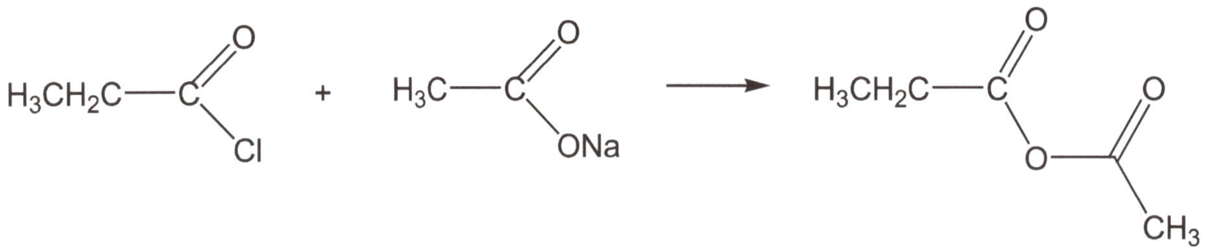

$$H_3CH_2C{-}COCl + H_3C{-}COONa \longrightarrow H_3CH_2C{-}CO{-}O{-}CO{-}CH_3$$

ethanoic propanoic anhydride

All the nucleophilic reactions shown with acyl chlorides can also be carried out using anhydrides. Acid anhydrides are more reactive than carboxylic esters but less reactive than acyl halides. Some typical reactions of acetic anhydride are shown in equations below. These equations show hydrolysis, alcoholysis, and aminolysis of anhydride bond. All these reactions break the anhydride bond and release carboxylic acid. The reaction with water gives two acid molecules. The reaction with alcohol gives an acid and an ester. The reaction with ammonia gives acid and an amide.

$$H_3C\text{-}C(=O)\text{-}O\text{-}C(=O)\text{-}CH_3 + H_2O \longrightarrow 2\ CH_3COOH$$

$$H_3C\text{-}C(=O)\text{-}O\text{-}C(=O)\text{-}CH_3 + R^1\text{-}OH \longrightarrow CH_3COOH + CH_3COOR^1$$

$$H_3C\text{-}C(=O)\text{-}O\text{-}C(=O)\text{-}CH_3 + NH_3 \longrightarrow CH_3CONH_2 + CH_3COOH$$

SUMMARY

Carboxylic acids are among the oldest known organic compounds. The functional group of the class is carboxylic group (COOH). Carboxylic acids are weaker acids than mineral acids but much stronger acids than alcohols and phenols because the carboxylate ions are stabilized by resonance. Carboxylic group has the highest priority among functional groups for nomenclature purposes. Carboxylic acids can be prepared by oxidation of alcohols and aldehydes, reactions of Grignard reagents with carbon dioxide, and acidic or alkaline hydrolysis of nitriles. Aromatic carboxylic acids can be prepared by oxidation of an alkyl chain having a benzylic hydrogen atom on the aromatic ring. Carboxylic acids can be transformed into a variety of carboxylic acid derivatives by the reaction at OH of the COOH functional group. The most common carboxylic acid derivatives are carboxylic esters, acid chlorides, acid anhydrides and amides.

Preparation of Carboxylic Acids

$$RCH_2OH \xrightarrow{CrO_3,\ H_3O^+} RCOOH \xleftarrow{CrO_3,\ H_3O^+} RCH{=}O$$

$$RMgX + CO_2 \xrightarrow{ether} R{-}\underset{\underset{O}{\|}}{C}{-}OMgX \xrightarrow{H_3O^+} R{-}\underset{\underset{O}{\|}}{C}{-}OH$$

$$RC{\equiv}N + 2\,H_2O \xrightarrow{H^+ \text{ or } HO^-} RCOOH + NH_3$$

$$ArCH_3 + KMnO_4 \longrightarrow ArCOOH$$

Reactions of Carboxylic Acids

$$RCOOH + H_2O \rightleftharpoons RCOO^- + H_3O^+ \quad \text{(ionization)}$$

$$RCOOH + NaOH \longrightarrow RCOO^-Na^+ + H_2O \quad \text{(formation of salts)}$$

$$RCOOH + R'OH \longrightarrow RCOOR' + H_2O \quad \text{(formation of esters)}$$

$$RCOOH + SOCl_2 \text{ or } PCl_5 \longrightarrow RCOCl \quad \text{(formation of acid chlorides)}$$

Reactions of Carboxylic Acid Derivatives

$$RCOOR' + NaOH \longrightarrow RCOO^-Na^+ + R'HO \quad \text{(saponification of esters)}$$

$$RCOOR' + NH_3 \longrightarrow RCONH_2 + R'OH \quad \text{(ammonolysis of esters amide formation)}$$

$$RCOOR' \xrightarrow[ether]{2\,R''MgX} R{-}\overset{R''}{\underset{R''}{C}}{-}OH + R'OH \quad \text{(reaction of esters with Grignard reagents)}$$

$$RCOOR' + LiAlH_4 \longrightarrow RCH_2OH + R'OH \quad \text{(reduction of esters)}$$

$$2\,R{-}CH_2COOR' \xrightarrow[2.\ H_3O^+]{1.\ R'O^-Na^+} R{-}\overset{H_2}{C}{-}\overset{\overset{O}{\|}}{C}{-}\underset{\underset{R}{|}}{\overset{H}{C}}{-}COOR' + R'{-}OH \quad \text{(Claisen condensation)}$$

$$RCONH_2 + H_2O \xrightarrow{H^+ \text{ or } HO^-} RCOOH + NH_3 \quad \text{(hydrolysis of amides)}$$

$$RCONH_2 + LiAlH_4 \longrightarrow RCH_2NH_2 \quad \text{(reduction of amides)}$$

RCOCl → (H_2O) RCOOH
RCOCl → (R'OH) RCOOR'
RCOCl → (NH_3) RCONH2

RC(O)OC(O)R' → (NH_3) $RCONH_2$
RC(O)OC(O)R' → (R"OH) RCOOR"

(nucleophilic substitution reactions of acid chlorides and anhydrides)

EXERCISES

1. Give the IUPAC name of the following compounds.

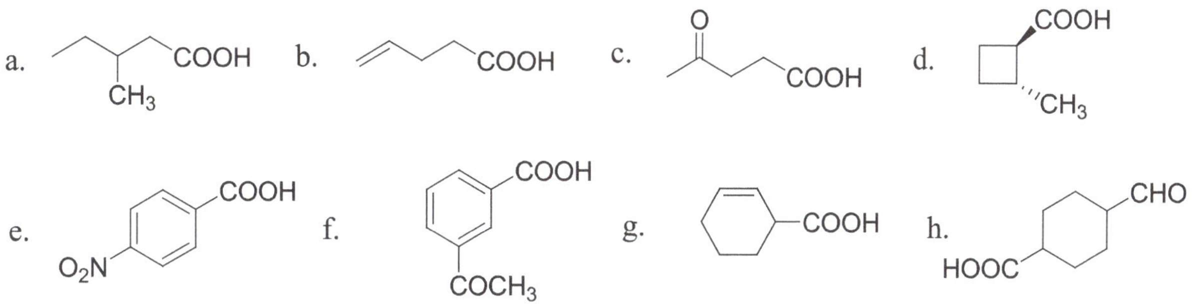

a. [structure: COOH, CH_3] b. [structure: COOH] c. [structure: O, COOH] d. [structure: COOH, CH_3]

e. [structure: O_2N, COOH] f. [structure: COOH, $COCH_3$] g. [structure: COOH] h. [structure: HOOC, CHO]

2. Draw a structural formula for the following carboxylic acids.
   a. 2-bromohexanoic acid
   b. methanoic acid
   c. 2-methylhex-4-enoic acid
   d. 3-bromo-4-chlorobenzoic acid
   e. 4-formylbenzoic acid
   f. cyclohept-2-enecarboxylic acid
   g. 3-isopropylcyclopentane carboxylic acid
   h. 3-oxohexanoic acid
   i. 5-methoxy-3-nitroheptanoic acid

3. Write equations to show at least three methods for the preparation of phenylacetic acid ($PhCH_2COOH$).

4. Show the stepwise reaction scheme for the synthesis of 4-isopropylbenzoic acid starting from benzene and other reagents.

5. Write the structure of carboxylic acids that will be obtained from the following reactions.

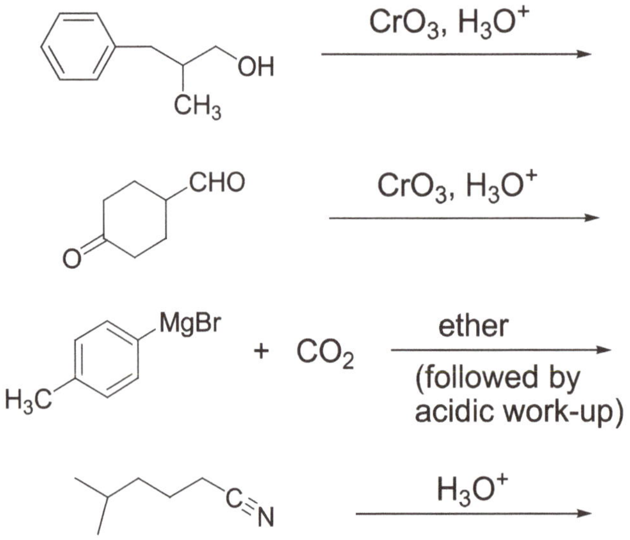

OH, CH_3 — CrO_3, H_3O^+ →

CHO — CrO_3, H_3O^+ →

MgBr, H_3C + CO_2 — ether → (followed by acidic work-up)

C≡N — H_3O^+ →

6. The pk_a of 2-chlorobutanoic acid is 2.85 and the pk_a of 3-chlorobutanoic acid is 4.05. Which one of the two is a stronger acid?

7. Arrange the compounds of each of the following series in order of increasing acidity (least acidic first).

a. 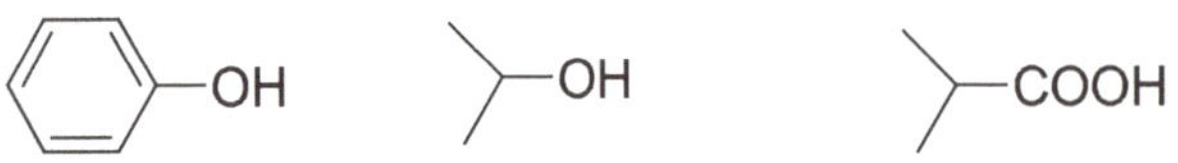 OH OH COOH

b. 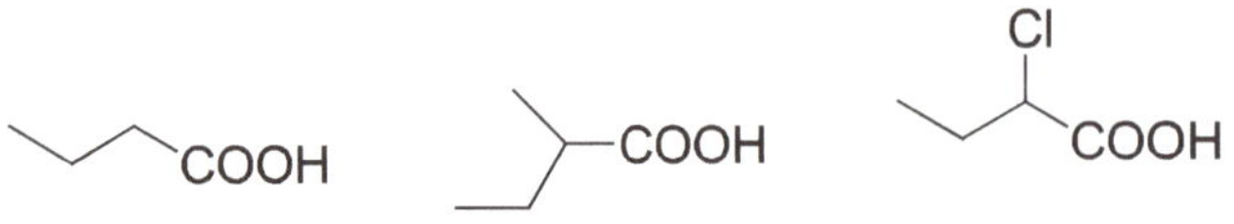 COOH COOH Cl COOH

c. 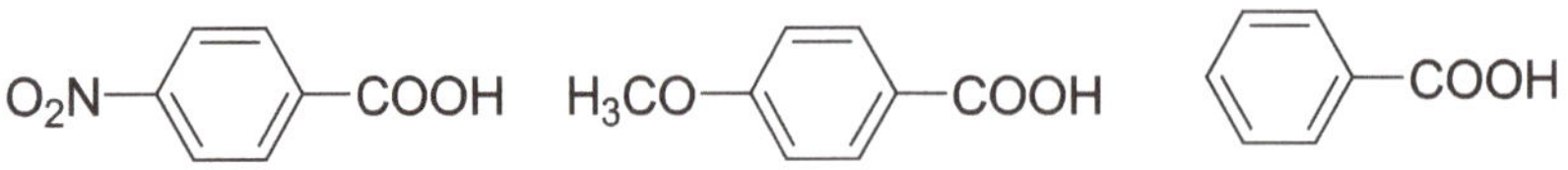 O_2N—COOH H_3CO—COOH COOH

8. Give the IUPAC name of the following compounds.

a. 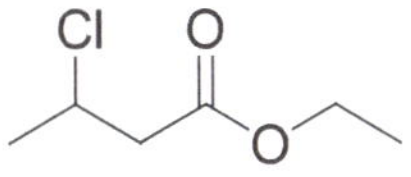 b. 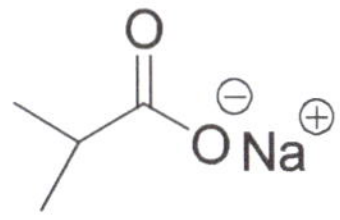 c. 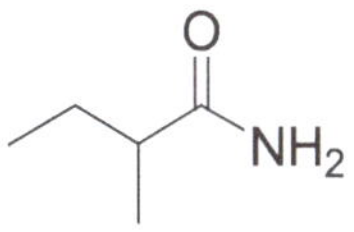 d.

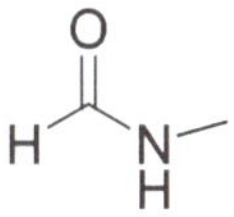

e. 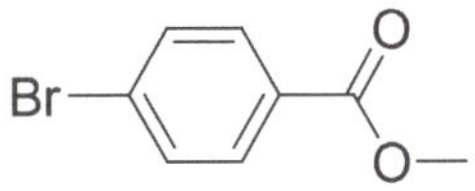 f. 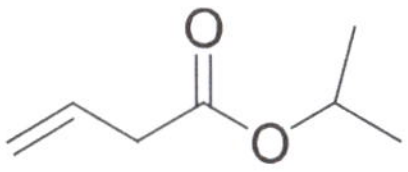 g. 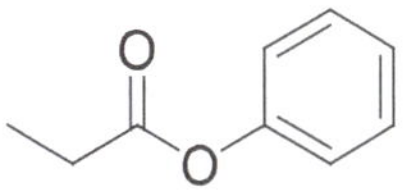 h.

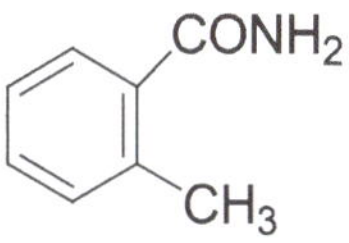

9. Draw a structural formula for the following carboxylic acid derivatives.

a.Calcium propanoate

b. ammonium benzoate

c. 4-phenylbenzoate

d. acetic anhydride

e. 3-ethoxybutanoyl chloride

f. *n*-butanamide

g. *N*-methylbenzamide

h. butyl 3-chlorohex-3-enoate

i. *p*-tolunitrile

j. ethyl formate

k. 4-methoxybenzoyl chloride

l. cyclohexyl acetate

10. Write out each step in the formation of ester from benzoic acid and ethanol.

11. Write an equation for the reaction of methyl benzoate with (a) hot aqueous potassium hydroxide, (b) ammonia on heating, and (c) ethylmagnesium bromide followed by an acidic treatment.

12. Show stepwise mechanism for (a) saponification of ethyl acetate, and (b) ammonolysis of ethyl acetate.

13. Identify the ester and the Grignard reagent that can be used in the preparation of following alcohols.

a. 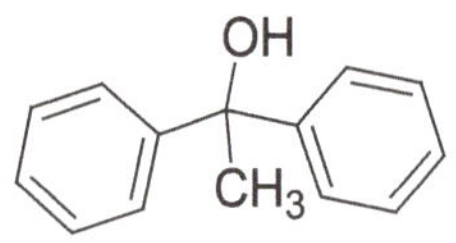 OH, CH_3

b. 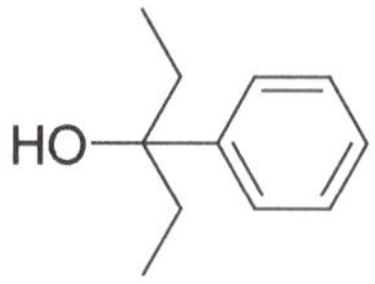 HO

14. Which of the two compounds in the following pairs will be more reactive towards nucleophile?

a. Ethyl acetate and acetone

b. ethyl acetate and acetyl chloride

15. Complete the following chemical equations by providing the product(s).

a.  COOH $\xrightarrow{SOCl_2}$

b.  + $KMnO_4$ $\longrightarrow$

c. 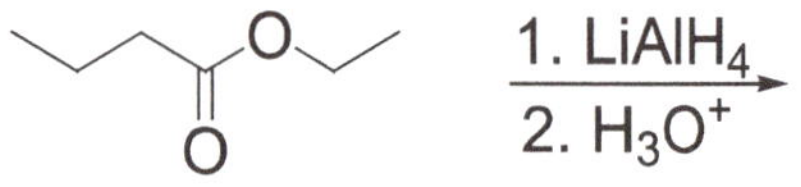 O, O $\xrightarrow[2.\ H_3O^+]{1.\ LiAlH_4}$

d. 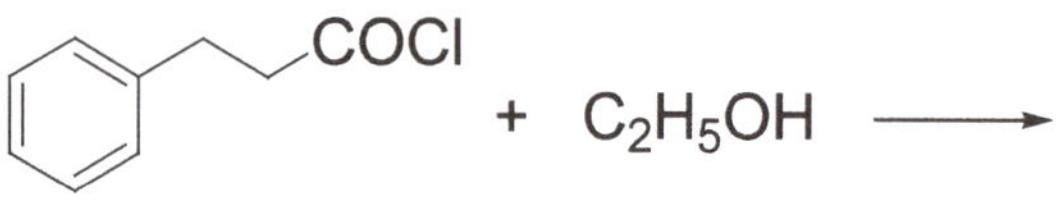 COCl + C_2H_5OH $\longrightarrow$

e. 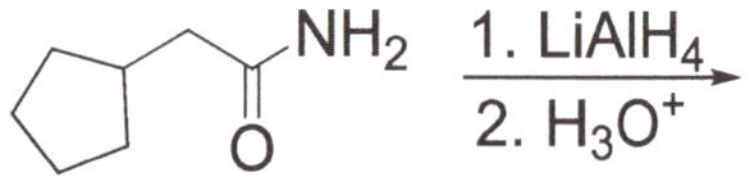 NH_2, O $\xrightarrow[2.\ H_3O^+]{1.\ LiAlH_4}$

f. 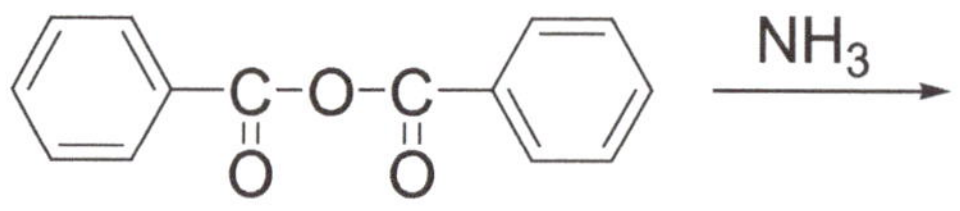 C-O-C, O, O $\xrightarrow{NH_3}$

16. Write a mechanism for each of the following reactions.

a. 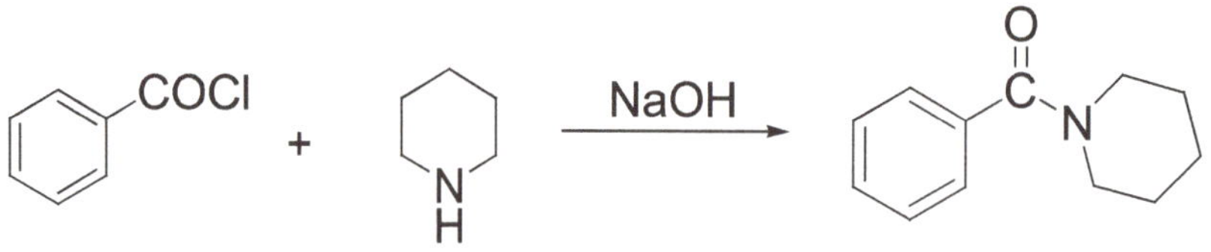 COCl + N H $\xrightarrow{NaOH}$ O, C, N

b. 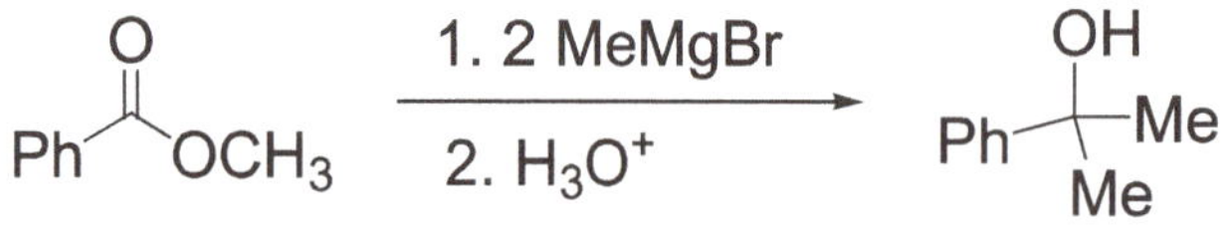 Ph(C=O)OCH_3 $\xrightarrow[2.\ H_3O^+]{1.\ 2\ MeMgBr}$ Ph–C(OH)(Me)–Me

17. The hydrolysis of an ester having molecular formula $C_8H_{16}O_2$ under acidic conditions gives an acid A and an alcohol B. Oxidation of alcohol B with Jone's reagent gave acid A. Identify the ester and write chemical equation for this reaction.

18. Write an equation for the following reactions.

a. reaction of butanoyl chloride with toluene in the presence of $AlCl_3$.

b. esterification of 1-butanol with acetic anhydride

c. hydrolysis of butanoyl bromide

d. ammonolysis of propanoyl chloride

e. hydrolysis of butanamide

f. hydrolysis of ethyl propanoate

19. Write the structure of the Claisen condensation product of the following esters.

a. Ethyl propanoate

b. ethyl 3-methylbutanoate

c. methyl phenylacetate

20. Show stepwise mechanism for the Claisen condensation of ethyl propanoate.

21. Which of the following esters can't undergo Claisen condensation and why?

a. O, Ph, OCH_3

b. O, OCH_3

c. $HCOOC_2H_5$

d. O, Ph, OC_2H_5

12 AMINES

Amines are considered a derivative of ammonia. They contain nitrogen atom attached to alkyl chains or non-aromatic or aromatic rings. In cyclic amines the nitrogen atom is part of the ring. Nitrogen atom in the amine has a lone pair of electrons; this makes amines basic in nature.

Several amines have been isolated from natural resources like plants. They play an important role in medicinal chemistry because many amines are biologically active. The well-known pain-killer morphine is an amine found in poppy seed. The cyclic amine nicotine is found in tobacco (**Figure 12.1**). A man-made compound 1,6-diaminohexane is used in textile industry in the synthesis of nylon.

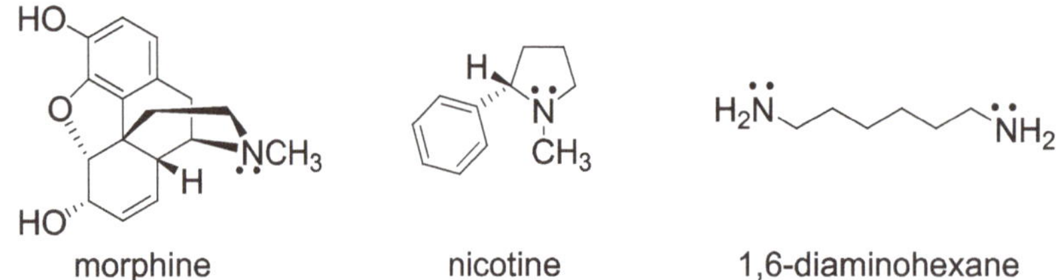

Fig.: 12.1: *Some amines of medicinal and industrial importance.*

12.1 Classification and Structure of Amines

Simple amines are organic bases derived from ammonia. The relationship between ammonia and amines can be illustrated by the following structures (**Figure 12.2**).

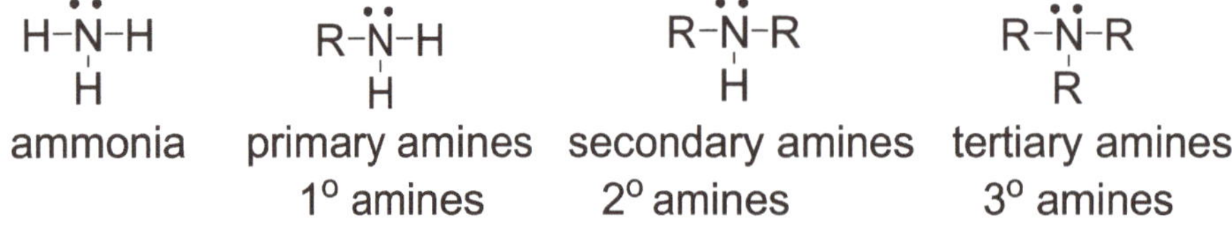

Fig. 12.2: *General structure of primary, secondary, and tertiary amines.*

Amines are classified as primary (1°), secondary (2°) and tertiary (3°) depending on whether one, two or three organic groups are attached to the nitrogen atom. The R group in these structures may be alkyl, cycloalkyl or aromatic, may be same or different. Secondary and tertiary amines can be cyclic as well in which the nitrogen atom will be present inside the ring (see the structure of morphine and nicotine).

The amines have a trivalent nitrogen atom which contains a lone pair of electrons. The nitrogen atom is therefore sp^3-hybridized, and the overall geometry is pyramidal (nearly tetrahedral).

12.2 Naming Amines

Amines are named in different ways. Simple amines are named by specifying the organic group attached to the nitrogen and adding suffix –amine (**Figure 12.3**).

NH_2 N H NH_2

n-butylamine diethylamine isopropylamine

Fig. 12.3: *Common name of simple amines.*

According to the IUPAC system, the amines can be named by either putting amine as suffix after removing "e" from the end of parent chain or amino group ($-NH_2$) can be named as a substituent, as shown in the following examples (**Figure 12.4**).

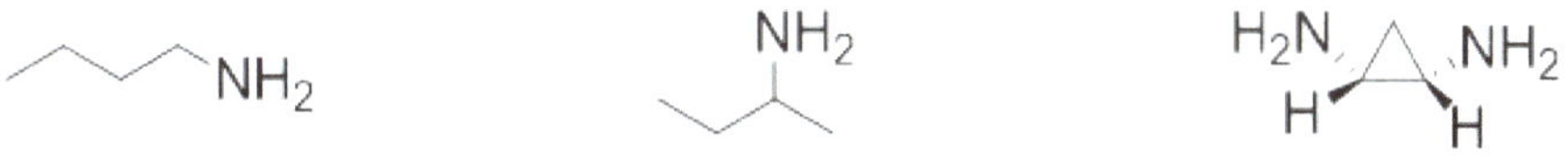

NH_2 NH_2 H_2N NH_2 H H

1-aminobutane or 1-butanamine

2-aminobutane or 2-butanamine

cis-1,2-diaminocyclopropane

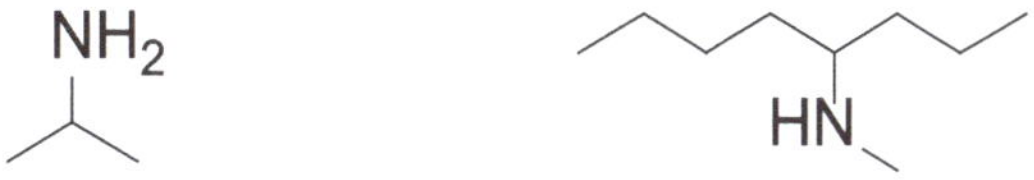

NH_2 HN

2-propanamine N-methyl-4-octanamine

Fig. 12.4: *IUPAC name of simple amines.*

As per this system, secondary or tertiary amines are named by using a prefix that includes everything except the longest possible chain (**Figure 12.5**).

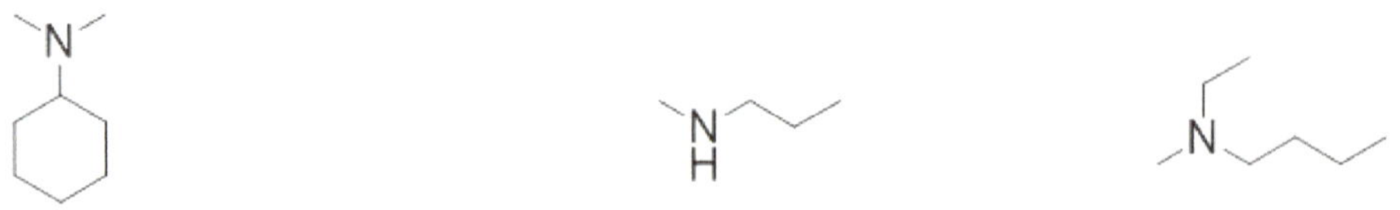

N,N-dimethylcyclohexanamine *N*-methylpropan-1-amine *N*-ethyl-*N*-methylbutan-1-amine

Fig. 12.5: *IUPAC name of secondary and tertiary amines.*

If the compound contains another functional group besides amino group, the amino group is named as a substituent (**Figure 12.6**).

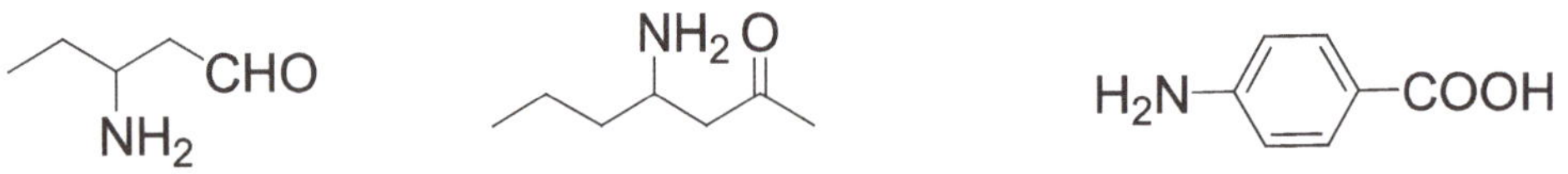

3-aminopentanal 4-aminoheptan-2-one 4-aminobenzoic acid

Fig. 12.6: *Naming amines with other functional groups.*

Aromatic amines are named either as derivatives of aniline (**Figure 12.7**) or by *Chemical Abstracts* names like alkyl amines. Aromatic amines are also popular by their common names.

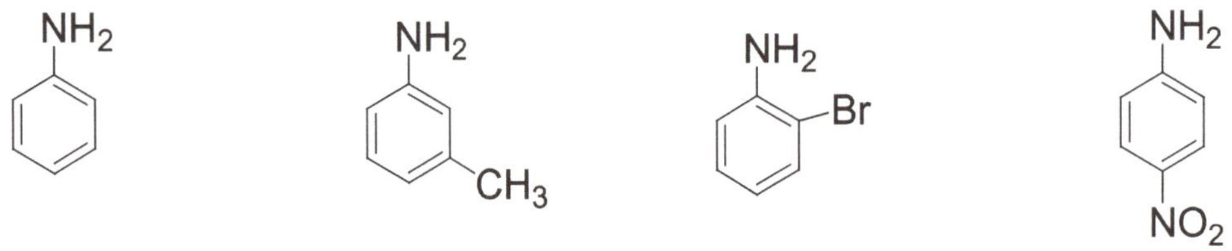

aniline	3-methylaniline	2-bromoaniline	4-nitroaniline
benzenamine	3-methylbenzenamine (*m*-toluidine)	2-bromobenzenamine	4-nitrobenzenamine

Fig. 12.7: *Naming aromatic amines.*

12.3 Basicity of Amines

The chemistry of amines is dominated by a lone pair of electrons on nitrogen atom. Due to the presence of a lone pair of electrons on nitrogen atom, amines are both basic and nucleophilic. Aqueous solutions of amines are basic because amines liberate hydroxide ions from water by abstracting a proton from it. $R\text{-}NH_3^+$ is conjugate acid of amine base.

$$R\text{-}\ddot{N}H_2 + H\text{-}OH \rightleftharpoons R\text{-}\overset{\oplus}{N}H_3 + \overset{\ominus}{O}H$$

The equilibrium constant for this reaction (see equation below) is called basicity constant, K_b. The K_b determines the ability of the amine to abstract the proton and hence to release the hydroxide ion. It is thus a measure of strength of a base, greater the K_b, stronger is the base. In terms of pK_b, lower the pK_b, stronger is the base (see **Table 12.1**)

$$K_b = \frac{[R\text{-}\overset{\oplus}{N}H_3][\overset{\ominus}{O}H]}{[R\text{-}NH_2]}$$

$$pK_b = -\log pK_b$$

It is evident from the table (Table 12.1) that the alkyl amines are more basic than ammonia itself. Alkyl groups are electron-donating groups. The presence of an electron-donating group on ammonium ion stabilizes the ammonium ion and thus reduces its acidity or increases its basicity. In general, the electron-donating groups increase the basicity of amines, and the electron-withdrawing groups decrease the basicity of amines. Furthermore, aromatic amines are much weaker base than alkyl or cycloalkyl amines. It is because of the delocalization of the lone pair on nitrogen to the aromatic ring (**Figure 12.8**). Due to delocalization, the electrons will be less available for an acid.

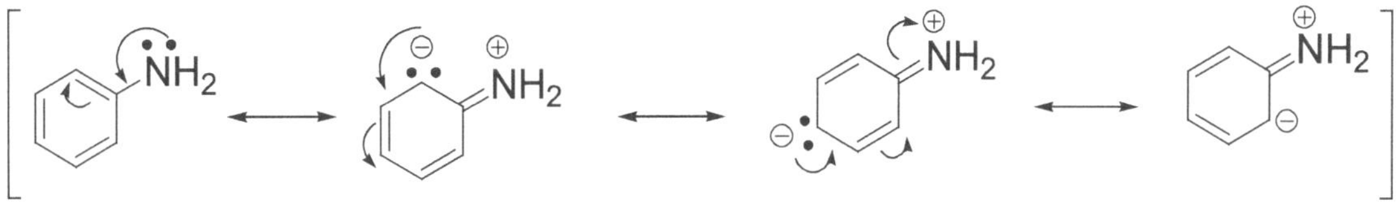

Fig. 12.8: *Delocalization of lone pair on nitrogen atom of aniline.*

Table 12.1: *Basicity of Some Common Amines*

Amines	**Formula**	**pK_b**	
Triethylamine	$(CH_3CH_2)_3N$	2.99	Most basic
Ethylamine	$CH_3CH_2NH_2$	3.19	
Dimethylamine	$(CH_3)_2NH$	3.27	
Methylamine	CH_3NH_2	3.34	
Diethylamine	$(CH_3CH_2)_2NH$	3.51	
Trimethylamine	$(CH_3)_3N$	4.19	
Ammonia	NH_3	4.74	Least basic

Problem: Which of the following two amines will be more basic and why?

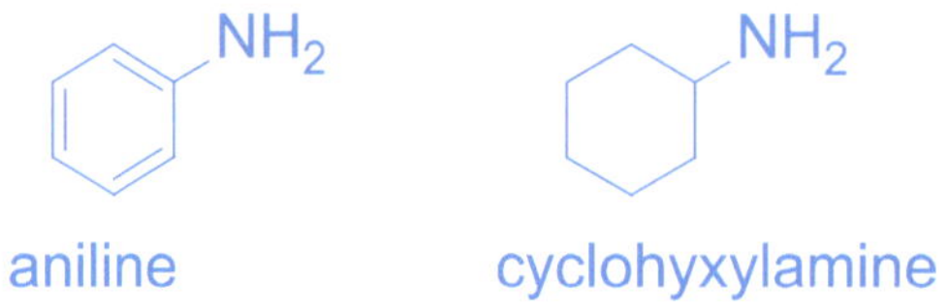

aniline cyclohyxylamine

Solution: Cyclohexylamine will be more basic than aniline. The lone pair of electrons in aniline is delocalized and hence less available for acids making aniline a weaker base in comparison to cyclohexylamine in which the lone pair on nitrogen atom is localized.

lone pair is localized on nitrogen atom

Comparing further the basicity of amines with amides, amides are non-basic; they do not react with acids. The aqueous solutions of amides are essentially neutral. In amides, there is a C=O group adjacent to nitrogen atom. So, the lone pair on nitrogen can delocalize on carbonyl group (**Figure 12.19**) and hence not available for acids.

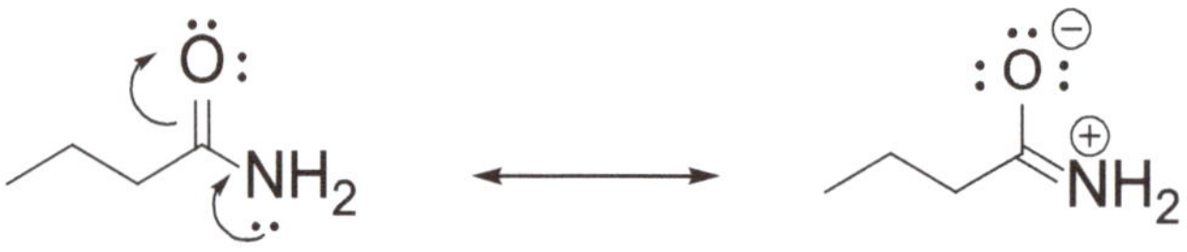

Fig. 12.9: *Delocalization of lone pair on nitrogen atom in amide.*

Problem: Arrange the following compounds in order of their increasing basicity.

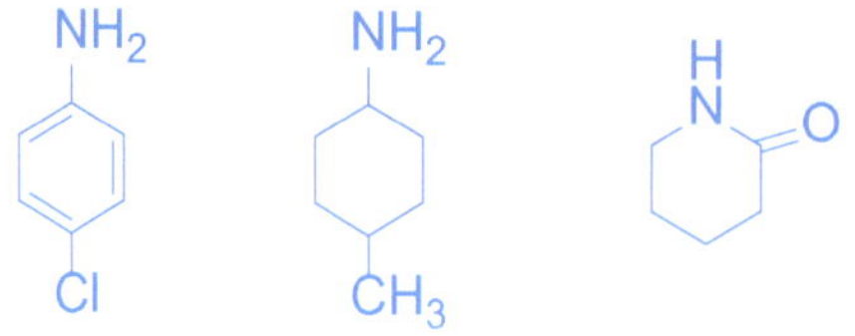

Solution:

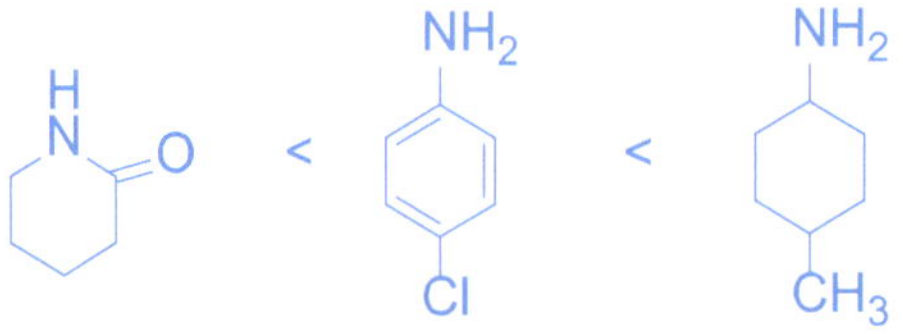

12.4 Preparation of Amines

There are many methods for the preparation of different classes of amines. You will notice that some of these methods, in fact, use the reactions of one amine to prepare another amine.

a. Reduction of Amides and Nitriles

Both amides and nitriles can be reduced with lithium aluminum hydride in ether solution to synthesize a primary amine. Nitriles, in turn, are prepared by nucleophilic substitution of alkyl halides with sodium cyanide. Both methods furnish amine with number of carbon atoms equal to the number of carbon atoms in the substrate. *N*-Substituted amides can be used for preparing secondary and tertiary amines.

$$R\text{-}C(=O)\text{-}NH_2 \xrightarrow[\text{2. } H_2O]{\text{1. } LiAlH_4\text{, ether}} R\text{-}CH_2\text{-}NH_2$$

$$R\text{-}X \xrightarrow{NaCN} \underset{\text{nitrile}}{R\text{-}C{\equiv}N} \xrightarrow[\text{2. } H_2O]{\text{1. } LiAlH_4\text{, ether}} R\text{-}CH_2\text{-}NH_2$$

b. Nucleophilic Substitution Reactions of Alkyl Halides with Ammonia or Amines

You have already seen the formation of primary, secondary, and tertiary amines in Section 7.2. Unfortunately, none of these reactions are of much synthetic value because it is difficult to stop the reaction at a particular stage and hence mixture of products is formed. For example, if ammonia is reacted with an alkyl halide it will give a primary alkyl amine that will further react with alkyl halide to give secondary amine. The process continues till all hydrogen atoms on nitrogen have been replaced with alkyl group forming a tertiary amine. Even this tertiary amine can react with alkyl halide to form an amine in which the nitrogen atom is positively charged. Such amines are known as quaternary salts.

$$R\text{-}X \xrightarrow{NH_3} R\text{-}NH_2 \xrightarrow{R\text{-}X} R_2NH \xrightarrow{R\text{-}X} R_2N\text{-}R \xrightarrow{R\text{-}X} \underset{\text{quaternary ammonium salt}}{R_4N^+X^-}$$

c. Reductive Amination of Aldehydes and Ketones

You have studied the formation of imines by addition of amines on C=O of aldehydes and ketones (Section 10.6). If the same reaction is carried out in the presence of a reducing agent such as lithium aluminum hydride, the imine formed undergoes reduction to give amine. This reaction known as reductive amination is an excellent method for the preparation of amines. A central nervous system (CNS) stimulant, amphetamine is prepared commercially by reaction of benzyl methyl ketone with ammonia using hydrogen gas over a nickel catalyst as a reducing agent.

$$RCHO \xrightarrow[\text{LiAlH}_4\text{, ether}]{R\text{-}NH_2} [RCH{=}N\text{-}R] \longrightarrow RCH_2\text{-}NH\text{-}R$$

$$C_6H_5CH_2COCH_3 \xrightarrow[H_2\text{-Ni}]{NH_3} C_6H_5CH_2CH(NH_2)CH_3$$

amphetamine: A CNS stimulant

d. Reduction of Nitro Group

Aromatic amines are synthesized by reduction of the nitro group in nitroarenes. For example, the nitro group on benzene ring is reduced by stannous chloride ($SnCl_2$) in aqueous hydrochloric acid. Treatment with base releases the free amine base.

$$C_6H_5NO_2 \xrightarrow[\text{2. NaOH}]{\text{1. SnCl}_2\text{-HCl (aq)}} C_6H_5NH_2$$

Problem: Write an equation for the preparation of each of the following amines.

a. Benzylamine ($PhCH_2NH_2$), b. *N*-benzylaniline ($PhNHCH_2Ph$), c. *p*-toluidine from benzene (Hint: more than one reaction required).

Solution:

$$C_6H_5C{\equiv}N \xrightarrow[\text{2. H}_3\text{O}^+]{\text{1. LiAlH}_4\text{, ether}} C_6H_5CH_2\text{-}NH_2$$

benzyl amine

$$C_6H_5\text{-}CHO + H_2N\text{-}C_6H_5 \xrightarrow[2.\ H_3O^+]{1.\ LiAlH_4,\ ether} C_6H_5\text{-}CH_2\text{-}NH\text{-}C_6H_5$$

N- benzylaniline

$$C_6H_6 \xrightarrow[AlCl_3]{CH_3Cl,} C_6H_5\text{-}CH_3 \xrightarrow{conc.\ HNO_3} p\text{-}CH_3C_6H_4NO_2 \xrightarrow[2.\ NaOH]{1.\ SnCl_2\text{-}HCl\ (aq.)} p\text{-}CH_3C_6H_4NH_2$$

p-toluidine

12.5 Reactions of Amines

You are by now already familiar with important reactions of amines – *N*-acylation of amines and *N*-alkylation of amines. The formation of amides from the reaction of carboxylic acid chlorides and amines (Section 11.7) is in fact *N*-acylation of amines – replacement of a hydrogen of amino group with an acyl group (R-C=O). The nucleophilic substitution of alkyl halides (Section 7.2) brings an alkyl group onto the nitrogen atom of amino group; hence it is called *N*-alkylation of amines. The succeeding section describes an important reaction of aromatic amines forming aromatic diazonium compounds.

12.6 Aromatic Diazonium Compounds

a. Preparation of diazonium Compounds

The reaction of primary aromatic amines with sodium nitrite and aqueous hydrochloric acid at 0-5 °C gives aryldiazonium chloride. The reaction is known as diazotization. Sodium nitrite reacts with hydrochloric acid to form nitrous acid (HONO) that reacts with amine to give the diazonium ion.

$$C_6H_5\text{-}NH_2 + NaNO_2 + HCl\ (aq) \xrightarrow[aqueous\ solution]{[HONO]\ 0\text{-}5\ ^{\circ}C} C_6H_5\text{-}N_2^+Cl^-$$

benzenediazonium chloride

$$Na^+NO_2^- + H^+Cl^- (aq) \xrightarrow[\text{aqueous solution}]{0\text{-}5\ ^oC} \underset{\text{nitrous acid}}{H\text{-}O\text{-}N{=}O} + NaCl$$

The mechanism involves a series of addition and elimination reactions in which one nitrogen atom from nitrosonium ion (NO), that is generated by protonation of nitrous acid, gets attached to amine nitrogen and two hydrogen atoms of amine are lost. Since two hydrogen atoms on amine nitrogen are required for this reaction, only primary aromatic amines can be diazotized.

$$O{=}N{-}OH + H^+ \rightleftharpoons O{=}N{-}\overset{\oplus}{O}H_2 \rightleftharpoons \underset{\text{nitrosonium ion}}{O{=}\overset{\oplus}{N}}$$

$$O{=}\overset{\oplus}{N} + Ph{-}NH_2 \longrightarrow Ph{-}\overset{\oplus}{N}H_2{-}N{=}O \rightleftharpoons Ph{-}NH{-}N{=}O + H^+$$

$$Ph{-}NH{-}N{=}O + H^+ \longrightarrow Ph{-}\overset{\oplus}{N}H{=}N{-}OH \longrightarrow \underset{\text{benzenediazonium ion}}{Ph{-}\overset{\oplus}{N}{\equiv}N}$$

b. Substitution Reactions: Applications in Organic Synthesis

Aromatic diazonium ions are valuable synthetic intermediates in organic chemistry. Several other functional groups can be introduced on the aromatic ring by replacing the diazonium ($-N_2^+$) group (see the scheme below). The aromatic diazonium ions are treated with Cu(I) chloride/bromide or cyanide to get aryl chloride/bromide or aryl cyanide. This reaction is known as the Sandmeyer reaction. Cyanide is easily converted to carboxylic acid. Aryl iodides and fluorides can also be prepared from diazonium ions using potassium iodide and tetrafluoroboric acid, respectively. Diazonium ions give phenols on heating in aqueous acid. These transformations have tremendous application in synthesis of diverse highly substituted benzene ring.

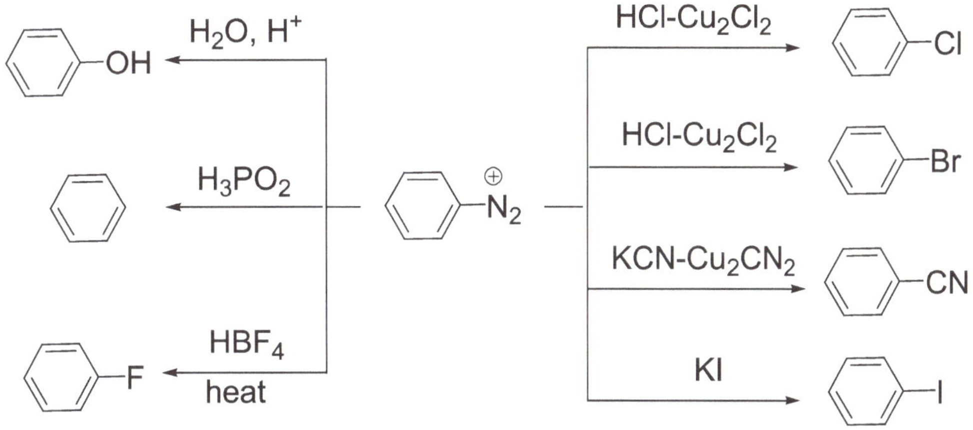

Problem: How can you synthesize *meta*-dichlorobenzene from benzene? Remember that chloro group on benzene ring is an *ortho, para*-directing group.

Solution: If you chlorinate benzene with excess chlorine, the major product will be *para*-dichlorobenzene because the chloro group is an *ortho,para*-directing group. In this situation, benzene will be first nitrated and then chlorinated to get *meta*-chloronitrobenzene. Reduction of the nitro group to an amino group followed by diazotization will generate *meta*-chlorobenzenediazonium ion which can be transformed to *meta*-dichlorobenzene by the Sandmeyer's reaction.

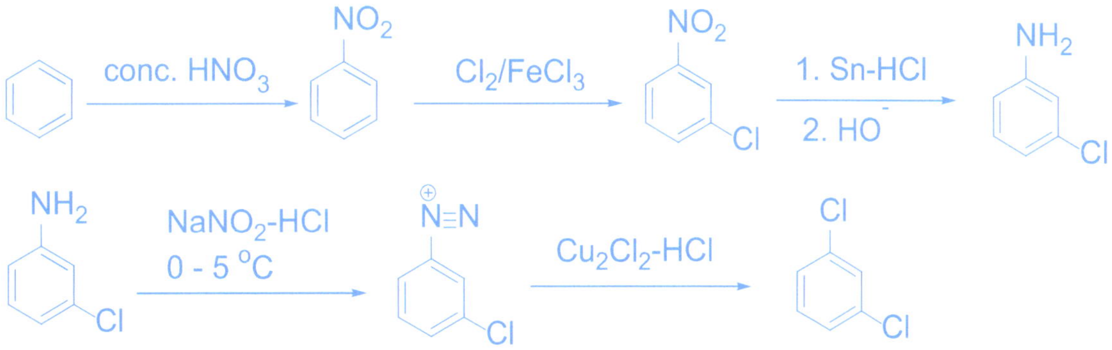

c. Diazo-coupling Reactions

Since diazonium ions carry a positive charge, they can react as an electrophile. However, they are very weak electrophile because the positive charge is delocalized to the aromatic ring. An aromatic ring carrying a strong activating group such as hydroxyl group or amino group undergoes electrophilic aromatic substitution with diazonium ions. This reaction is known as diazo-coupling because both nitrogen atoms are retained in the product as –N=N-. Usually, the reaction occurs at the *para*-position of the ring but if the para-position is occupied, the reaction takes place at *ortho*-position. The product *diazo* compounds are highly colored and used as dye in textile industry.

$C_6H_5N_2^+ + C_6H_5OH \xrightarrow{HO^-} C_6H_5{-}N{=}N{-}C_6H_4{-}OH$

p-hydroxyazobenzene

12.7 Heterocyclic Amines

You have already seen examples of cyclic ethers, cyclic esters, and cyclic amides. All such compounds come into the category of heterocyclic compounds. Heterocyclic compounds are such compounds in which the ring consists of one or more "heteroatom", the atom other than carbon atom. Heterocyclic compounds occur frequently in natural resources such as plants, marines, and other microorganisms. Many of the heterocyclic compounds have been discovered to have different kind of medicinal properties. A great deal of research work in organic chemistry is therefore devoted to the discovery of novel heterocyclic compounds and their chemistry.

Heterocyclic amines containing one or more nitrogen atoms in the ring are also found frequently in nature and have been utilized in past for treating some diseases. Many of these amines are known with their special name but they can be named in similar way as alkanes by putting the *prefix* "aza" before the name of cycloalkyl ring. For example, a three-membered ring with a nitrogen atom in the ring is known as aziridine but it can be named as azacyclopropane (propane because the ring is three-membered). If there is one or more substituent on the ring, locate the position of the substituent by counting from nitrogen atom.

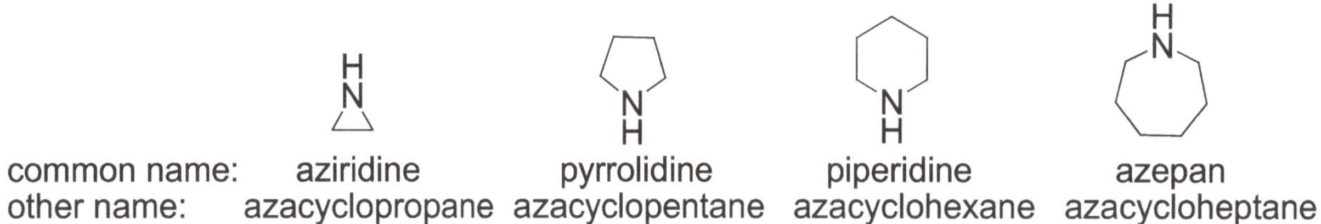

The chemistry of heterocyclic amines is like their open-chain counterparts. However, they show some unique properties if the ring is unsaturated. Let's take here the examples of pyrrole and pyridine. Vitamins B_6 present in yeast, cereals, and other foodstuffs, and necessary for the synthesis of some amino acids are substituted pyridines.

a. Pyrrole

Pyrrole has a five-membered ring containing one nitrogen atom. The ring is unsaturated because it has two carbon-carbon double bonds. Pyrrole is an aromatic compound. Even though there are only five atoms in the ring it has six π electrons in a cyclic conjugated system just like benzene. Each of the four carbon atoms in the ring contributes one π electron and the *sp*^{2}-hybridized nitrogen atom contributes its lone pair of electrons making pyrrole an aromatic ring system of 6π electrons in the planar ring. As a result of aromatization, pyrrole can undergo electrophilic aromatic substitution reactions. However, pyrrole does not show the typical basic properties of amines because the lone pair of electrons on nitrogen atom is involved in π system and no more available for an acid.

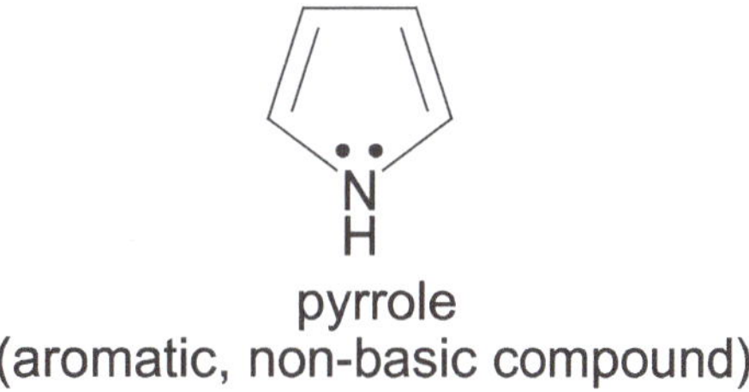

pyrrole
(aromatic, non-basic compound)

b. Pyridine

Pyridine has a six-membered ring containing one nitrogen atom and three carbon-carbon double bonds. It is, thus, nitrogen analogue of benzene. Like benzene, it is a flat molecule with bond angles of 120°. It is also a 6π system with each carbon and *sp*^{2}-hybridized nitrogen atom contributing one electron.

Unlike pyrrole, however, pyridine is a basic (pK_b = 8.75) molecule. The lone pair of electrons on nitrogen atom in pyridine is not involved in π system and hence is available for reaction with an acid.

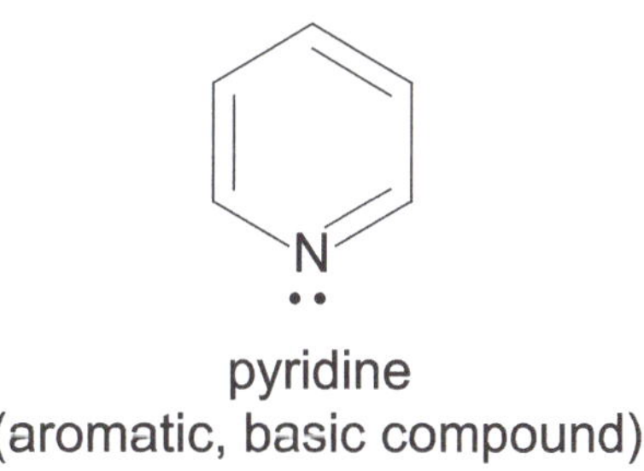

pyridine
(aromatic, basic compound)

SUMMARY

Amines are nitrogen containing compounds. Amines occur frequently in plants. Many amines of medicinal value have been isolated from plants. Amines can be classified as primary (RNH2), secondary (R2NH), and tertiary amines (R3N).

Amines are weak bases. Alkyl amines are stronger base than ammonia because of electron donating effect of the alkyl group. Alkyl amines are also stronger base in comparison to aromatic amines because the lone pair of electrons in the latter case is delocalized to aromatic ring and hence not available for proton.

Alkyl amines can be prepared by nucleophilic substitution of alkyl halides but superior methods are available. The reduction of amides and nitriles with lithium aluminum hydride affords amines. Reductive amination of aldehydes and ketones serves as a good method for the preparation of secondary amines. Aromatic amines can be prepared by reduction of nitro group on the aromatic ring using stannous chloride and hydrochloric acid.

$$R{-}C({=}O){-}NH_2 \xrightarrow[\text{2. } H_2O]{\text{1. } LiAlH_4\text{, ether}} R{-}CH_2{-}NH_2$$

$$R{-}X \xrightarrow{NaCN} \underset{\text{nitrile}}{R{-}C{\equiv}N} \xrightarrow[\text{2. } H_2O]{\text{1. } LiAlH_4\text{, ether}} R{-}CH_2{-}NH_2$$

$$R(H)C{=}O \xrightarrow[LiAlH_4\text{, ether}]{R{-}NH_2} \left[R(H)C{=}N{-}R \right] \longrightarrow R(H)CH{-}\overset{H}{N}{-}R$$

$$C_6H_5{-}CH_2{-}C({=}O){-}CH_3 \xrightarrow[H_2\text{-}Ni]{NH_3} \underset{\text{amphetamine}}{C_6H_5{-}CH_2{-}CH(NH_2){-}CH_3}$$

$$C_6H_5{-}NO_2 \xrightarrow[\text{2. NaOH}]{\text{1. } SnCl_2\text{-HCl (aq)}} C_6H_5{-}NH_2$$

Amines undergo *N*-acylation and *N*-alkylation on reaction with acyl chlorides and alkyl halides, respectively. Aromatic amines are oxidized by nitrous acid to aromatic diazonium ions which serve as a powerful synthetic reagent. Diazonium ion on aromatic ring can be replaced with different functionalities using appropriate reagents. Diazonium ions also undergo diazo-coupling reaction with aromatic rings containing strong ring-activating group. The diazo coupling takes place at *para* position of the ring, but if the *para* position is occupied then diazo group gets attached to *ortho* position.

$$RCOCl + R'NH_2 \longrightarrow RCONHR' + HCl$$

$$RCl + R'NH_2 \longrightarrow R'NHR + HCl$$

$$ArNH_2 + NaNO_2 + HCl \xrightarrow{0\text{-}5^oC} ArN_2^+Cl^-$$

EXERCISES

1. Write a structural formula for each of the following compounds containing an amine functional group.

   a. tributylamine b. 3-aminohexane c. 3-methyl-2-pentanamine

   d. isopropylamine e. *m*-chloroaniline f. 4-bromoaniline

   g. *N,N*-dimehtylaminocyclopentane h. 1,4-diaminobutane

   i. tetramethylammonium chloride j. *p*-toluidine k. *sec*-butylamine

2. Give correct name to each of the following structures of amines.

a. 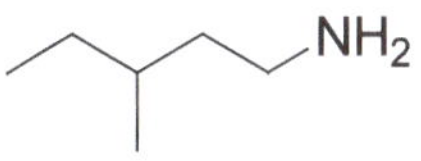 NH_2 b. 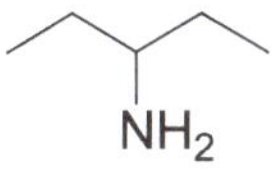 NH_2 c. 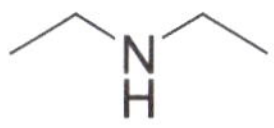 N H d. 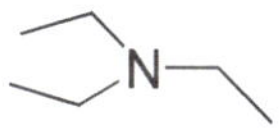 N

e. 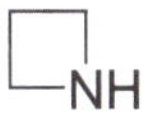 NH f. 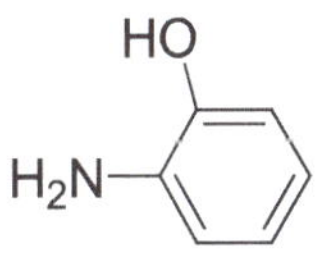 HO H_2N g. NH_2 CH_3 h. NH_2

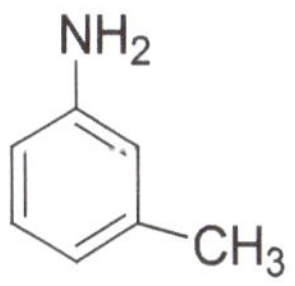

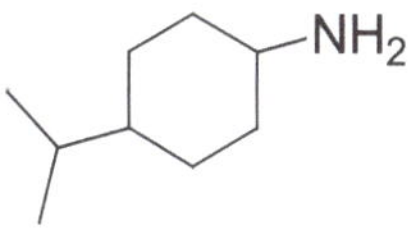

3. Identify the following amines as primary (1°), secondary (2°), tertiary (3°), and quaternary ammonium salt.

a.  b. 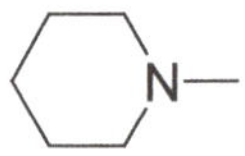 N— c. 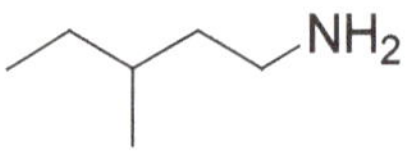 NH2 d. 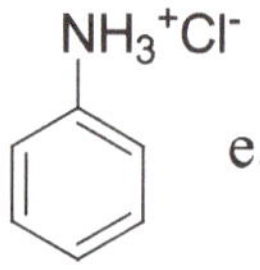 $NH_3^+Cl^-$ e. 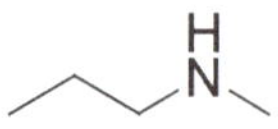 H N f. $(CH_3)_3N$

4. Give the structures of amides required for the preparation of following amines.

a. 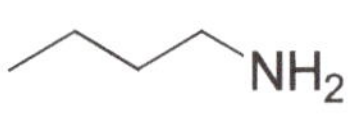 NH_2 b. 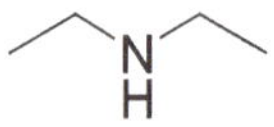 N H c. 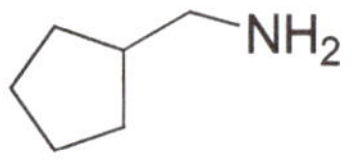 NH_2 d. 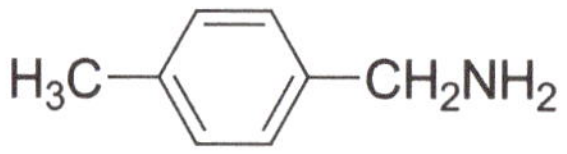 H_3C—CH_2NH_2

5. Give the structures of nitriles, which on reduction, will give the following amines.

a. 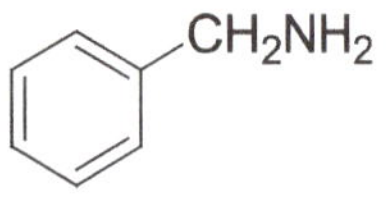 CH_2NH_2 b. 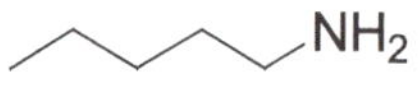 NH_2 c. 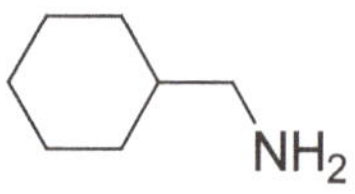 NH_2

6. Reduction of nitriles is a good method for the preparation of primary amines. Can the following primary amine be prepared from a nitrile? If not, what is the reason?

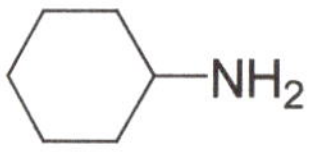

—NH_2

7. Show the reaction equation for the preparation of following amines by reductive amination.

a. 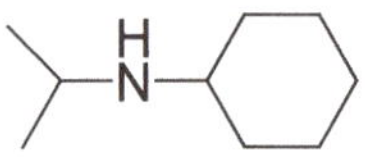 H N b. 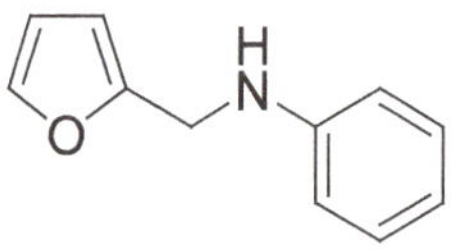 O H N c. 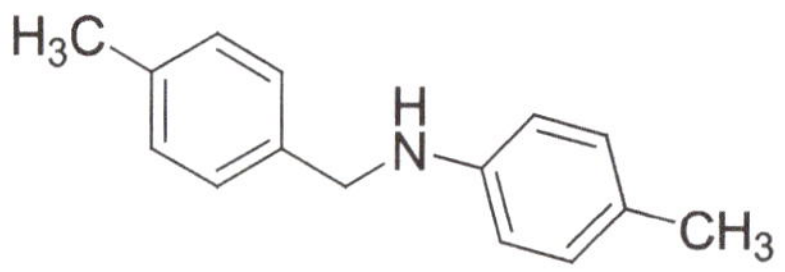 H_3C H N CH_3

8. Show a reaction scheme for a synthesis of each of the following amines assuming there is no amide, nitrile, and aldehyde or ketone is present in the laboratory.

a. 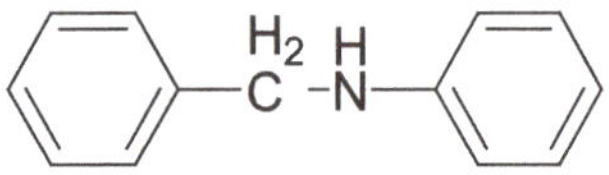 H_2 H C N b. 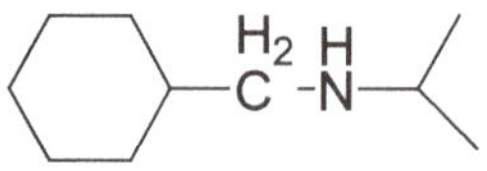 H_2 H C N c. CH_2NH_2

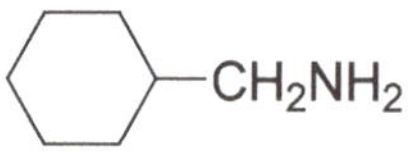

9. Show a reaction scheme for a synthesis of each of the following aromatic amines from benzene or toluene.

a. 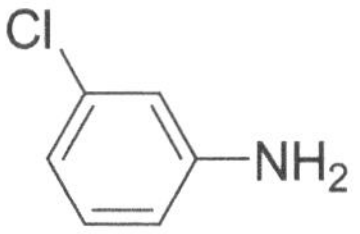 Cl, $-NH_2$ b.  H_3C- $-NH_2$ c. 2,4,6-tribromoaniline d. 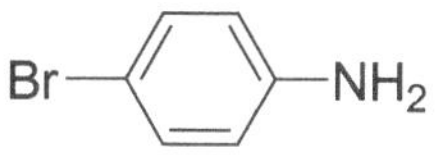 $Br-$ $-NH_2$

10. Which compound in each of the following pair is more basic and why?

a. sodium hydroxide and ethylamine

b. propylamine and propanamide

c. diethylamine and ethylphenylamine

d. diethylamine and triethylamine

11. Arrange the compounds of each group in order of increasing basicity.

i) a. $CH_3CH_2NH_2$ b. $CH_3CH_2CONH_2$ c. $CH_3CH_2NHCH_2CH_3$.

ii) a. 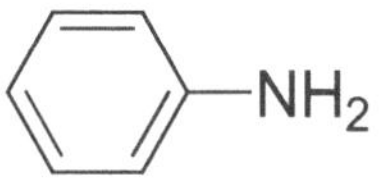 $-NH_2$ b. 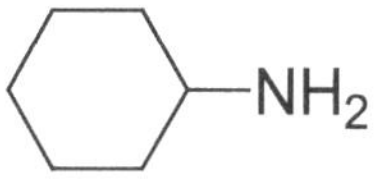 $-NH_2$ c. 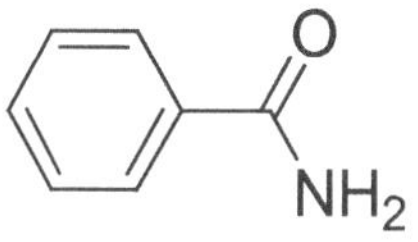 O NH_2

iii) a. 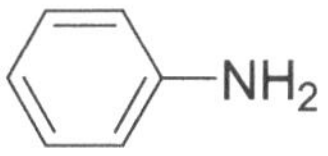 $-NH_2$ b. 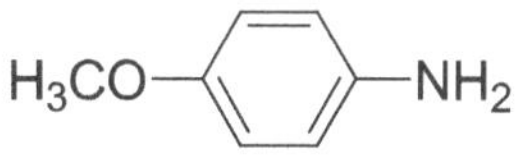 H_3CO- $-NH_2$ c.  O_2N- $-NH_2$

12. How can you explain the following order of basicity?

$(CH_3)_2NH$ (pK_b 3.28) > CH_3NH_2 (pK_b 3.38) > $(CH_3)_3N$ (pK_b 4.28) > NH_3 (pK_b 4.76)

13. How can you separate a mixture of phenol, aniline, and acetophenone?

14. Write the structure of the product in the following reaction.

a. $CH_3CH_2NH_2$ + HCl ⟶

b. $(CH_3CH_2)_3N$ + HI ⟶

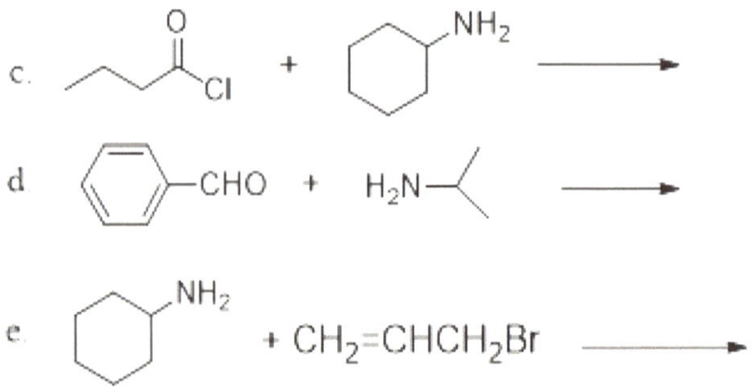

c. O, Cl + NH_2 ⟶

d. CHO + H_2N ⟶

e. NH_2 + $CH_2{=}CHCH_2Br$ ⟶

15. Write an equation for the preparation of *p*-toluenediazonium chloride from *p*-toluidine.

16. Write equations for the reaction of *p*-toluenediazonium chloride with the following reagents under given conditions.

a. Hypophosphorous acid b. potassium iodide

c. heating with HBF_4 d. KCN and cuprous cyanide

e. HCl and cuprous chloride f. heating with an aqueous acid

g. 4-methoxyphenol and HO^-.

17. Show the application of diazonium ion in synthesizing-----

a. *m*-nitrophenol from *m*-nitroaniline.

b. 4-bromobenzoic acid from 4-bromoaniline.

c. *m*-chlorobenzonitrile from benzene.

d. benzene to *p*-chlorobenzonitrile.

APPENDIX

Priority of Functional Groups in Nomenclature

(Highest to Lowest)

1. Carboxylic acids (COOH)
2. Sulfonic acids (SO_3H)
3. Esters (RCOOR)
4. Acyl halides (RCOX)
5. Amides ($CONH_2$)
6. Nitriles (CN)
7. Aldehydes (CHO)
8. Ketones ($R_2C=O$)
9. Alcohols (R-OH)
10. Amines (NH_2)

Priority in Alkenes and Alkynes

If a carbon-carbon double and carbon-carbon triple bonds are present in a molecule and are equidistant in the parent chain, carbon-carbon double bond will have priority in nomenclature (C=C will get lower number).

Bond Lengths in Organic Compounds

C–H	length (Å)	C–C	length (Å)	multiple-bonds	length (Å)
sp3–H	1.10	sp^3–sp^3	1.54	benzene	1.40
sp2–H	1.09	sp^3–sp^2	1.50	alkenes	1.34
sp–H	1.08	sp^3–sp	1.46	alkynes	1.20
		sp^2–sp	1.43	allenes (C=C=C)	1.30

Average Carbon-Carbon Bond Strength Based on Hybridization

Carbon-carbon bonds	Ethane H_3C-CH_3	Ethylene H_2C=CH_2	Acetylene HC≡CH
Hybridization	sp^3- sp^3	sp^2-sp^2	sp-sp
Bond strength	88 kcal/mol	152 kcal/mol	200 kcal/mol

p*K*a of Some Common Compounds

Compound	Formula	pKa
Methanoic acid	HCOOH	3.75
Acetic acid	CH_3COOH	4.75
Benzoic acid	C_6H_5COOH	4.20
4-Nitrobenzoic acid	4-$O_2NC_6H_4COOH$	3.41
4-Methoxybenzoic acid	4-$H_3COC_6H_4COOH$	4.47
Oxalic acid	HOOC-COOH	1.2
Phenol	C_6H_5OH	10
p-anisole	p-$H_3COC_6H_4OH$	10.21

Compound	**Formula**	**pKa**
4-nitrophenol	4-$O_2NC_6H_4OH$	7.15
Ethanol	C_2H_5OH	15.9
Acetone/2-propanone	CH_3COCH_3	19
Ethane	C_2H_6	50
Ethylene	C_2H_4	44
Acetylene	C_2H_2	25
Hydrocyanic acid	HCN	9.30
Sulfuric acid	H_2SO_4 HSO_4^-	-3 1.99
Hydrochloric acid	HCl	-7
Nitric acid	HNO_3	-1.4
Water	H_2O	14
Ammonia	NH_3	33
Alkyl amines	R-NH_2	Around 40
Cyclohexanamine	$C_6H_{11}NH_2$	42
Ethanamide	$CH_3CH_2CONH_2$	15

PERIODIC TABLE OF THE ELEMENTS

1	2	3	4	5	6	7	8	9	10	11	12	13	14	15	16	17	18
1 **H** 1.00794																	2 **He** 4.00260
3 **Li** 6.941	4 **Be** 9.01218											5 **B** 10.81	6 **C** 12.011	7 **N** 14.0067	8 **O** 15.9994	9 **F** 18.9984	10 **Ne** 20.179
11 **Na** 22.9898	12 **Mg** 24.305											13 **Al** 26.9815	14 **Si** 28.0855	15 **P** 30.9738	16 **S** 32.06	17 **Cl** 35.453	18 **Ar** 39.948
19 **K** 39.0983	20 **Ca** 40.08	21 **Sc** 44.9559	22 **Ti** 47.88	23 **V** 50.9415	24 **Cr** 51.996	25 **Mn** 54.9380	26 **Fe** 55.847	27 **Co** 58.9332	28 **Ni** 58.69	29 **Cu** 63.546	30 **Zn** 65.38	31 **Ga**	32 **Ge** 72.59	33 **As** 74.9216	34 **Se** 78.96	35 **Br** 79.904	36 **Kr** 83.8
37 **Rb** 85.4678	38 **Sr** 87.62	39 **Y** 88.9059	40 **Zr** 91.22	41 **Nb** 92.9064	42 **Mo** 95.94	43 **Tc** (98)	44 **Ru** 101.07	45 **Rh** 102.906	46 **Pd** 106.42	47 **Ag** 107.868	48 **Cd** 112.41	49 **In** 114.82	50 **Sn** 118.69	51 **Sb** 121.75	52 **Te** 127.6	53 **I** 126.9	54 **Xe** 131.29
55 **Cs** 132.905	56 **Ba** 137.33	71 **Lu** 174.967	72 **Hf** 178.49	73 **Ta** 180.948	74 **W** 183.85	75 **Re** 186.207	76 **Os** 190.2	77 **Ir** 192.22	78 **Pt** 195.08	79 **Au** 196.967	80 **Hg** 200.59	81 **Tl** 204.383	82 **Pb** 207.2	83 **Bi** 208.908	84 **Po** (209)	85 **At** (210)	86 **Rn** (222)
87 **Fr** (223)	88 **Ra** 226.025	103 **Lr** (260)	104 **Rf** (261)	105 **Db** (262)	106 **Sg** (263)	107 **Bh** (264)	108 **Hs** (265)	109 **Mt** (268)	110 Uun (269)	111 Uuu (272)	112 Uub (269)		114 Uuq		116 Uuh		118 Uuo

Lanthanides:

57 **La** 138.906	58 **Ce** 140.12	59 **Pr** 140.908	60 **Nd** 144.24	61 **Pm** (145)	62 **Sm** 150.36	63 **Eu** 151.96	64 **Gd** 157.25	65 **Tb** 158.925	66 **Dy** 162.50	67 **Ho** 161.930	68 **Er** 167.26	69 **Tm** 166.934	70 **Yb** 173.04

Actinides:

89 **Ac** 227.028	90 **Th** 232.038	91 **Pa** 231.036	92 **U** 238.029	93 **Np** 237.048	94 **Pu** (244)	95 **Am** (243)	96 **Cm** (247)	97 **Bk** (247)	98 **Cf** (251)	99 **Es** (252)	100 **Fm** (257)	101 **Md** (258)	102 **No** (259)

INDEX

P

Q

R

S

www.ingramcontent.com/pod-product-compliance
Lightning Source LLC
LaVergne TN
LVHW071139160826
845679LV00003B/510

* 9 7 9 8 8 9 1 8 6 5 4 6 4 *